Mellin-Transform Method for Integral Evaluation
Introduction and Applications to Electromagnetics

Mellin-Transform Method for Integral Evaluation

George Fikioris

www.morganclaypool.com

ISBN: 978-3-031-00569-5   paperback
ISBN: 978-3-031-00569-5   paperback

ISBN: 978-3-031-01697-4   ebook
ISBN: 978-3-031-01697-4   ebook

DOI: 10.1007/978-3-031-01697-4

A Publication in the Springer series
SYNTHESIS LECTURES ON COMPUTATIONAL ELECTROMAGNETICS #13

Lecture #13
Series Editor: Constantine A. Balanis, Arizona State University

Series ISSN: 1932-1252   print
Series ISSN: 1932-1716   electronic

First Edition
10 9 8 7 6 5 4 3 2 1

Mellin-Transform Method for Integral Evaluation

George Fikioris
School of Electrical and Computer Engineering
National Technical University of Athens
Athens, Greece

SYNTHESIS LECTURES ON COMPUTATIONAL ELECTROMAGNETICS #13

ABSTRACT

This book introduces the Mellin-transform method for the exact calculation of one-dimensional definite integrals, and illustrates the application if this method to electromagnetics problems. Once the basics have been mastered, one quickly realizes that the method is extremely powerful, often yielding closed-form expressions very difficult to come up with other methods or to deduce from the usual tables of integrals. Yet, as opposed to other methods, the present method is very straightforward to apply; it usually requires laborious calculations, but little ingenuity. Two functions, the generalized hypergeometric function and the Meijer G-function, are very much related to the Mellin-transform method and arise frequently when the method is applied. Because these functions can be automatically handled by modern numerical routines, they are now much more useful than they were in the past.

The Mellin-transform method and the two aforementioned functions are discussed first. Then the method is applied in three examples to obtain results, which, at least in the antenna/electromagnetics literature, are believed to be new. In the first example, a closed-form expression, as a generalized hypergeometric function, is obtained for the power radiated by a constant-current circular-loop antenna. The second example concerns the admittance of a 2-D slot antenna. In both these examples, the exact closed-form expressions are applied to improve upon existing formulas in standard antenna textbooks. In the third example, a very simple expression for an integral arising in recent, unpublished studies of unbounded, biaxially anisotropic media is derived. Additional examples are also briefly discussed.

KEYWORDS

Integration (Mathematics), Mellin Transforms, Antenna Theory, Electromagnetic Theory

To Yannis, Vassilis and little Nefeli

Contents

Preface

This book introduces the Mellin-transform method for the exact calculation of one-dimensional definite integrals and illustrates the application of this method to electromagnetics problems. Once the basics have been mastered, one quickly realizes that the method is extremely powerful, often yielding closed-form expressions very difficult to come up with other methods or to deduce from the usual tables of integrals. Yet, as opposed to other methods, the present method is very straightforward to apply; it usually requires laborious calculations, but little ingenuity. Two functions, the generalized hypergeometric function and the Meijer G-function, are very much related to the Mellin-transform method and arise frequently when the method is applied. Because these functions can be automatically handled by modern numerical routines, they are now much more useful than they were in the past.

The Mellin-transform method and the two aforementioned functions are discussed first. Then the method is applied in three examples to obtain results which, at least in the antenna/electromagnetics literature, are believed to be new. In the first example, a closed-form expression, as a generalized hypergeometric function, is obtained for the power radiated by a constant-current circular-loop antenna. The second example concerns the admittance of a two-dimensional slot antenna. In both these examples, the exact closed-form expressions are applied to improve upon existing formulas in standard antenna textbooks. In the third example, a very simple expression for an integral arising in recent, unpublished studies of unbounded, biaxially anisotropic media is derived. Additional examples are also briefly discussed.

Parts of the text have been used in a graduate course on mathematical methods in electromagnetic theory at the author's university. The book proceeds from first principles and is suitable for self-study. The reader is assumed to be familiar with elementary complex analysis, including the idea of analytic continuation; otherwise, the text is self-contained. The book is an expanded version of a review paper entitled "Integral Evaluation Using the Mellin Transform and Generalized Hypergeometric Functions: Tutorial and Applications to Antenna Problems," published in the December 2006 issue of the *IEEE Transactions on Antennas and Propagation* [1]. Much of Sections 2.1–2.7 came from a seminar written by the author and Professor Dionisios Margetis, and presented in 1996 at the Air Force Research Laboratory, Sensors Directorate, where the author was then working. Finally, the author is greatly indebted to Professor Tai Tsun Wu for originally introducing him to Mellin transforms.

CHAPTER 1

INTRODUCTION

A quick look through any advanced textbook on electromagnetics or antennas reveals the important role played by definite integrals. In studies of radiation problems, workers in the field integrate along wire antennas, over planar apertures, and over enclosing surfaces; they obtain exact solutions to differential equations using integral transforms such as Fourier and Laplace transforms and solve integrodifferential equations approximately using projection methods, such as Galerkin's method, which introduce additional integrations. More often than not, the integrals encountered are complicated and one resorts to numerical–integration techniques. But investigating whether the integrals can be evaluated analytically is always worth some effort: Closed-form expressions are usually preferable for numerical calculations, especially when the expressions involve special functions computable by packaged routines. Furthermore, such expressions can be useful for further analytical work and can provide better physical insight.

This book reviews a technique for evaluating one-dimensional definite integrals exactly. Its power is evidenced by the fact that it is used by modern packages that perform symbolic integration. In particular, it forms an important part of Mathematica's* routine `Integrate[ ]` which, according to S. Wolfram [2], "can evaluate (. . .) most definite integrals listed in standard books of tables." Furthermore, the technique "has been used in an essential manner" [3] for the creation of the integral tables in the monumental, three-volume reference work [4]–[6] by A. P. Prudnikov, Yu. A. Brychkov, and O. I. Marichev. We call the technique the "Mellin-transform method," because taking a Mellin transform is the method's initial and key step. But it is known in the literature [2] with other names, such as the Marichev–Adamchik method [3], [7]–[9].

A key feature of the Mellin-transform method is that it often provides results in terms of generalized hypergeometric functions ($_pF_q$) or, more generally, in terms of Meijer G-functions or Mellin–Barnes integrals. The $_pF_q$, which are defined by convergent series, can often be rewritten in terms of simpler special functions; thus, the $_pF_q$ are frequently—but not always—just a convenient intermediate step. The same is true for the G-function, whose definition is more involved and intimately related to the Mellin transform. However, even

---

* Mathematica is a registered product of Wolfram Research, Inc., Champaign, IL 61820-7237, USA.

expressions involving $_pF_q$ can nowadays be very useful for numerical calculations: For the numerical computation of $_pF_q$, today's packaged routines use sophisticated methods and do not rely exclusively on the aforementioned series definition. Such routines—which will, hopefully, further improve in the near future—can be used as black boxes by the user. Today, packaged routines exist even for the more general G-function.

Why take the trouble to learn the Mellin-transform method? Why not just use modern symbolic integration routines? Given an integral, one can (and, in the author's opinion, should!) first attempt evaluation with Mathematica or other packages. One should also try lookup in standard integral tables. Nevertheless, as this book will explicate, learning the method is worthwhile for a number of reasons: (i) The method is (once the basics have been mastered) easy to apply. (ii) We can sometimes combine the method with additional manipulations to yield further useful results. (iii) Learning the method serves as an excellent introduction to the $_pF_q$ and even more so to the G-function. Thus, familiarity with the method can help us appreciate and understand our answers. (iv) Intermediate expressions (especially the expression involving a Mellin–Barnes integral) can greatly help further analytical work including, in particular, asymptotic analysis. Thus sometimes one is interested in more than just the final result. (v) Finally, many workers (the author included!) simply like to check their results independently. This is especially true when the integral to be evaluated contains several parameters upon which the form of the answer depends.

It is worth mentioning that, even in the primarily mathematical literature, the Mellin-transform method is often underutilized. For example, it is barely mentioned in D. Zwillinger's 1992 *Handbook of Integration* [10]. As another example, when discussing the technique (in the very much related context of asymptotic expansions), M. J. Ablowitz and A. S. Fokas [11, p. 504] state, "This method, although often quick and easy to apply, is not widely known."

A major portion of this book (Chapters 2–4) is a general description of the Mellin-transform method, concluding (Section 4.2) with a first example of the method's application. The discussion in Chapters 2–4 proceeds from first principles and includes a review of the gamma function, the Mellin transform, the $_pF_q$, the G-function, and Mellin–Barnes integrals. Our treatment here is introductory: We pay little attention to generality or to the many mathematical subtleties of our subject. In particular, we do not provide validity conditions for formulas involving general functions. Chapters 2–4 are, necessarily, not very different from other introductory treatments of the subject [9], [12]–[14]. Some elementary complex analysis, including the idea of analytic continuation, is a prerequisite for understanding Chapters 2–4, which, otherwise, are self-contained.

The next two chapters (5 and 6) present two original example-integrals, arising in antenna problems (loop, microstrip, and aperture antennas), to which the Mellin-transform method is applied. In each example, we give background information (theory and/or applications), use the Mellin-transform method to evaluate the relevant integrals exactly, and discuss, apply, or interpret the exact results. Chapter 7 presents a similar discussion for a certain integral arising in unpublished studies on biaxially anisotropic media; our treatment here combines the Mellin-transform method with additional manipulations and is more advanced.

Chapters 8 and 9 and Appendices B and C present further material, including a brief comment on the related topic of asymptotic expansions, an examination of alternative methods that can yield certain of our results, a discussion of relations of our exact results to entries in standard integral tables, and a guide to literature related to the Mellin-transform method. Appendix A is a stand-alone discussion on the convergence of definite integrals, giving simple rules that can help one determine whether a given integral converges or diverges; these rules are used often throughout this book. Finally, Appendix D gives two additional examples of the application of the Mellin-transform method to electromagnetics problems.

CHAPTER 2

Mellin Transforms and the Gamma Function

2.1 MELLIN TRANSFORM: DEFINITION, STRIP OF ANALYTICITY (SOA)

Let $f(x)$ denote a complex-valued function of the real, positive variable x. The Mellin transform of $f(x)$ will be denoted by $\tilde{f}(z)$ and, alternatively, by the more complete notation $\text{MT}\{f(x); z\}$. The *definition of the Mellin transform* involves an integral

$$\tilde{f}(z) = \text{MT}\{f(x); z\} = \int_0^\infty x^{z-1} f(x)\, dx. \tag{2.1}$$

The new variable z, which is taken to be complex, must be restricted to those values for which the integral converges. In general, we have convergence at $x = 0$ only if $\text{Re}\{z\}$ is larger than a certain value and at $x = \infty$ only if $\text{Re}\{z\}$ is smaller than a certain value. This is readily understood from the results of Appendix A, which is stand-alone discussion on the convergence of integrals.

Thus, if the Mellin transform of $f(x)$ (as defined in (2.1)) exists at all, it exists in a vertical strip in the complex z-plane. In some cases, the strip reduces to a half plane. Furthermore, under mild conditions on $f(x)$, it can be shown [7, p. 39] that $\tilde{f}(z)$ is an analytic function of z in that strip. The strip will be referred to by the term "strip of analyticity" (SOA). Although usual in the literature, the term SOA is somewhat misleading because analytic continuation of $\tilde{f}(z)$ to other complex values of z is usually possible. In fact, for the application considered here, analytic continuation is always necessary. (A more accurate term for the SOA would perhaps be "strip of initial definition.")

2.2 MELLIN TRANSFORM: BASIC PROPERTIES

We now turn to some properties of the Mellin transform. First of all, it is related to other, more usual, transforms. For example, if

$$\text{FT}\{f(x); z\} = \int_{-\infty}^\infty f(x)\, e^{ixz}\, dx, \tag{2.2}$$

is the Fourier transform of $f(x)$, one has

$$\tilde{f}(z) = \text{FT}\{f(e^x); -iz\}. \tag{2.3}$$

Also, if $\text{LT}\{f(x); z\}$ is the usual (one-sided) Laplace transform of $f(x)$, then

$$\tilde{f}(z) = \text{LT}\{f(e^{-x}); z\} + \text{LT}\{f(e^x); -z\}. \tag{2.4}$$

Thus, many properties of the Laplace and Fourier transforms can be rephrased for the Mellin transform. For instance, the aforementioned analyticity of the Mellin transform in a vertical strip can be viewed as a consequence of the well-known analyticity of the Laplace transform in a right-half plane.

It can be shown (via the Fourier or Laplace inversion formula) that the *inversion formula for the Mellin transform* is [7, p. 39], [12]

$$f(x) = \frac{1}{2\pi i} \int_{\delta - i\infty}^{\delta + i\infty} x^{-z} \tilde{f}(z) \, dz, \tag{2.5}$$

where the integration path is a vertical line in the complex z-plane, *lying within the SOA,* as shown in Fig. 2.1. Formula (2.5) uniquely determines $f(x)$ from $\tilde{f}(z)$.

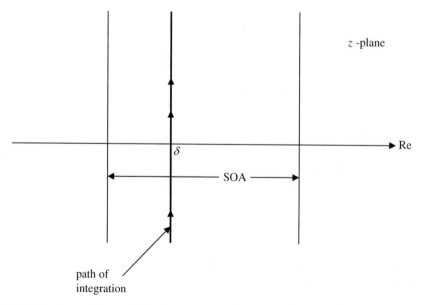

FIGURE 2.1: The integration path in the inversion formula (2.5) is a vertical line in the complex z-plane, lying within the strip of analyticity SOA. *(Figure adapted from Fikioris [1])*

Let $\delta_L < \text{Re}\{z\} < \delta_R$ be the SOA of $\tilde{f}(z)$. The reader is invited to show, directly from the definition (2.1), the following elementary properties of the Mellin transform:

$$\text{MT}\{f(\alpha x); z\} = \alpha^{-z}\tilde{f}(z), \qquad \delta_L < \text{Re}\{z\} < \delta_R, \quad \alpha > 0, \tag{2.6}$$

$$\text{MT}\{x^\alpha f(x); z\} = \tilde{f}(z+\alpha), \qquad \delta_L - \text{Re}\{\alpha\} < \text{Re}\{z\} < \delta_R - \text{Re}\{\alpha\}, \tag{2.7}$$

and

$$\text{MT}\{f(x^\alpha); z\} = \frac{1}{|\alpha|}\tilde{f}\left(\frac{z}{\alpha}\right), \qquad \begin{cases} \alpha\delta_L < \text{Re}\{z\} < \alpha\delta_R, & \text{if } \alpha > 0, \\ \alpha\delta_R < \text{Re}\{z\} < \alpha\delta_L, & \text{if } \alpha < 0. \end{cases} \tag{2.8}$$

2.3   MELLIN TRANSFORM: PARSEVAL FORMULA AND RELATED PROPERTIES

The Fourier or Laplace transform of the *product* of two functions is given by the *convolution* of the individual transforms (where convolution is defined differently for the two transforms). The corresponding statement for the Mellin transform is

$$\int_0^\infty g(y)h(y)y^{z-1}\,dy = \frac{1}{2\pi i}\int_{\delta-i\infty}^{\delta+i\infty} \tilde{g}(w)\tilde{h}(z-w)\,dw, \tag{2.9}$$

in which δ belongs to the SOA of $\tilde{g}(w)$. Once again, the right-hand side is a convolution of sorts. To show (2.9), begin from its left-hand side, use (2.5) to introduce $\tilde{g}(w)$, and interchange the order of integration. Given the SOAs of $\tilde{g}(z)$ and $\tilde{h}(z)$, one can readily determine the vertical strip (in the complex z-plane) in which (2.9) holds.

A slight generalization of (2.9) is

$$\int_0^\infty g(xy)h(y)y^{z-1}\,dy = \frac{1}{2\pi i}\int_{\delta-i\infty}^{\delta+i\infty} \tilde{g}(w)\tilde{h}(z-w)x^{-w}\,dw, \qquad x > 0, \tag{2.10}$$

which is a combination of (2.9) and (2.6). The special case $z = 1$ of (2.10) is

$$\int_0^\infty g(xy)h(y)\,dy = \frac{1}{2\pi i}\int_{\delta-i\infty}^{\delta+i\infty} \tilde{g}(z)\tilde{h}(1-z)x^{-z}\,dz, \tag{2.11}$$

in which δ belongs both to the SOA of $\tilde{g}(z)$ and to the SOA of $\tilde{h}(1-z)$—for (2.11) to hold, it is necessary that these two SOAs overlap. Note that the special case $x = 1$ of (2.11) is what is usually called the *Parseval formula for the Mellin transform*.

Formula (2.11) forms the core of the Mellin-transform method. It is worth restating (2.11) somewhat differently, and outlining an alternative derivation. The operation on the left-hand side is the so-called *Mellin convolution* of the two functions $g(x)$ and $h(x)$; we (but not all authors—see Section 9.1 in our discussion in Chapter 9) denote it by $(g \oslash h)(x)$ and so, by definition,

$$(g \oslash h)(x) = \int_0^\infty g(xy)h(y)\,dy, \quad x > 0. \tag{2.12}$$

The fundamental difference from the more usual types of convolution is that the product xy, not the difference $x - y$, is the argument of one of the two integrand functions. By virtue of the inversion formula, we can rewrite (2.11) and (2.12) as

$$\mathrm{MT}\{(g \oslash h)(x); z\} = \tilde{g}(z)\tilde{h}(1 - z), \tag{2.13}$$

which can also be shown directly from (2.12), (2.6), and (2.1), with no recourse to an inversion formula. In the right-hand side of (2.13) we have a product of Mellin transforms (one of them is actually reflected and translated; see the relevant remarks in Section 9.1), and so formula (2.13) is, in a certain sense, the reverse of (2.9), with the *Mellin convolution* in the original (x-) domain and a *product* in the transform (z-) domain.

2.4   GAMMA FUNCTION

The gamma function $\Gamma(z)$ is defined as the Mellin transform of e^{-x}. We routinely use $\Gamma(z)$ when applying the Mellin-transform method. We have

$$\Gamma(z) = \int_0^\infty x^{z-1} e^{-x}\,dx, \qquad \mathrm{Re}\{z\} > 0. \tag{2.14}$$

By Rule 1 of Appendix A, the restriction $\mathrm{Re}\{z\} > 0$ in (2.14) is a necessary and sufficient condition for convergence of the integral at $x = 0$. The restriction means that the SOA is, in this case, the entire right-half complex z-plane. It can be shown that

$$\Gamma(z) = \sum_{n=0}^\infty \frac{(-1)^n}{n!} \frac{1}{z+n} + \int_1^\infty x^{z-1} e^{-x}\,dx, \qquad z \neq 0, -1, -2, \dots \tag{2.15}$$

(split (2.14) into $\int_0^1 + \int_1^\infty$; expand e^{-x} into Taylor series in first integral; integrate term by term). In the derivation of (2.15), it was assumed that $\mathrm{Re}\{z\} > 0$. However, the right-hand-side of (2.15) is analytic for all z except $z = 0, -1, -2, \dots$. Thus, (2.15) provides the analytic continuation of $\Gamma(z)$ to (complex) values of z for which the defining integral (2.14) did not make sense.

The gamma function has many properties. We give the ones most useful for the Mellin-transform method. From (2.15), it is seen that at $z = 0, -1, -2, \dots, -n, \dots, \Gamma(z)$ has *simple poles* with corresponding *residues* $1, -1, \frac{1}{2}, \dots, \frac{(-1)^n}{n!}, \dots$. The *recurrence formula*

$$\Gamma(z+1) = z\Gamma(z) \tag{2.16}$$

is easily shown by integrating (2.14) by parts. With $\Gamma(1) = 1$, it follows that

$$\Gamma(n+1) = n!, \quad n = 0, 1, 2, \ldots. \tag{2.17}$$

It is possible to show [15] that

$$\Gamma(z)\Gamma(1-z) = \frac{\pi}{\sin \pi z}, \tag{2.18}$$

which is called the *reflection formula*. As a consequence of (2.18), $\Gamma(\frac{1}{2}) = \sqrt{\pi}$. With the recurrence formula (2.16), one can further determine the values $\Gamma(\frac{3}{2}), \Gamma(\frac{5}{2}), \ldots$. As an additional consequence of (2.18), $1/\Gamma(z)$ is analytic in the entire z-plane. That is, $\Gamma(z)$ has *no zeros*.

The *duplication formula* [15] is

$$\Gamma(2z) = \frac{1}{2\sqrt{\pi}} 2^{2z} \Gamma(z) \Gamma\left(z + \frac{1}{2}\right), \tag{2.19}$$

which can be generalized for $\Gamma(3z), \Gamma(4z), \ldots$ by the *multiplication formula* [15]

$$\Gamma(nz) = \frac{1}{\sqrt{n(2\pi)^{n-1}}} \, n^{nz} \prod_{l=0}^{n-1} \Gamma\left(z + \frac{l}{n}\right), \qquad n = 1, 2, \ldots. \tag{2.20}$$

In the right-hand side of (2.20), note that all coefficients of z in the arguments of the gamma functions are 1. Finally, the familiar *Stirling's formula* [15]

$$\Gamma(z) = \sqrt{\frac{2\pi}{z}} \left(\frac{z}{e}\right)^z \left[1 + O\left(\frac{1}{z}\right)\right], \quad \text{as } |z| \to \infty \text{ with } |\arg z| < \pi, \tag{2.21}$$

is an asymptotic approximation to $\Gamma(z)$ for large, complex arguments. To obtain such an approximation valid for $\text{Re}\{z\} < 0$, combine (2.21) with the reflection formula (2.18).

A plot of $\Gamma(z)$ and $1/\Gamma(z)$ for the case of *real* z is given in Fig. 2.2. Observe the rapid increase (decrease) of $\Gamma(z)$ ($1/\Gamma(z)$) for large, positive z in accordance with (2.21) and (2.17) as well as the poles (zeros) of $\Gamma(z)$ ($1/\Gamma(z)$) at $z = 0, -1, -2, -3, -4$.

2.5   PSI FUNCTION

In the Mellin-transform method, we often encounter the derivative $\Gamma'(z)$ of $\Gamma(z)$. This is usually computed from $\Gamma(z)$ and the *psi function* $\psi(z)$, defined by

$$\psi(z) = \frac{\Gamma'(z)}{\Gamma(z)}. \tag{2.22}$$

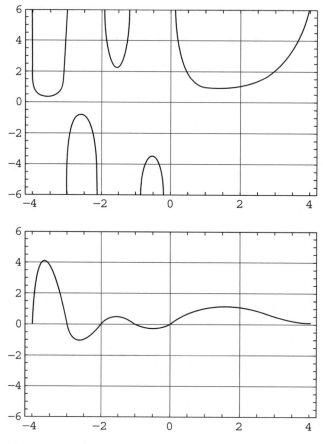

FIGURE 2.2: The functions $\Gamma(z)$ (top) and $1/\Gamma(z)$ (bottom) as functions of z (z real, with $-4 < z < 4$)

The recurrence formula for the psi function, which is a simple consequence of the recurrence formula (2.16), is

$$\psi(z+1) = \frac{1}{z} + \psi(z). \qquad (2.23)$$

It follows from (2.23) that

$$\psi(n+1) = \psi(1) + \left(\frac{1}{1} + \frac{1}{2} + \cdots + \frac{1}{n}\right), \quad n = 1, 2, \ldots. \qquad (2.24)$$

The value $\psi(1)$ can be calculated [15] from

$$-\psi(1) = \gamma = 0.5772156649\ldots, \qquad (2.25)$$

where γ is Euler's constant. Equation (2.24) then allows calculation of $\psi(2), \psi(3), \ldots$.

2.6   POCHHAMMER'S SYMBOL

The definition of *Pochhammer's symbol* $(z)_n$ is

$$(z)_n = \frac{\Gamma(z+n)}{\Gamma(z)}, \qquad n = 0, 1, 2, \ldots. \qquad (2.26)$$

Pochhamer's symbol will be used when defining the $_pF_q$. It satisfies the recurrence formula

$$(z)_{n+1} = (z+n)(z)_n, \qquad n = 0, 1, 2, \ldots. \qquad (2.27)$$

2.7   SIMPLE APPLICATIONS

To familiarize readers with the foregoing material and to prepare for what follows, we now provide simple examples involving the Mellin transform and $\Gamma(z)$, $\psi(z)$, and $(z)_n$.

Application 1: Let us calculate $\mathrm{MT}\{e^{-x} - 1; z\}$ and directly verify the inversion formula.

With $f(x) = e^{-x} - 1$, the integrand of (2.1) behaves like x^z as $x \to 0$ and like x^{z-1} as $x \to \infty$. Therefore, by Rules 1 and 3 of Appendix A, the integral converges if $-1 < \mathrm{Re}\{z\} < 0$. For such values of z, it is seen that

$$\mathrm{MT}\{e^{-x} - 1; z\} = \int_0^\infty (e^{-x} - 1)x^{z-1}\, dx = \Gamma(z), \quad -1 < \mathrm{Re}\{z\} < 0 \qquad (2.28)$$

(integrate by parts; identify resulting integral with $\Gamma(z+1)$; use recurrence formula (2.16)). Thus, the Mellin transforms of e^{-x} and $e^{-x} - 1$ are both $\Gamma(z)$. The two SOAs, however, do not overlap. Therefore, the two corresponding inversion formulas (2.5) are different because δ changes.

To directly verify that

$$\frac{1}{2\pi i} \int_{\delta-i\infty}^{\delta+i\infty} x^{-z}\Gamma(z)\, dz = e^{-x} - 1, \quad -1 < \delta < 0 \qquad (2.29)$$

(compare this with the inversion formula for $\Gamma(z)$), collapse the contour until it wraps around the poles at $z = -1, -2, \ldots$ on the negative real axis—a procedure to be described, somewhat loosely, as "closing the contour at left." By (2.21) and (2.18) (see also Chapter 8), this procedure is indeed possible so that, by the residue theorem,

$$\frac{1}{2\pi i} \int_{\delta-i\infty}^{\delta+i\infty} x^{-z}\Gamma(z)\, dz = \sum_{n=1}^\infty \mathrm{Res}\{x^{-z}\Gamma(z); z = -n\} = \sum_{n=1}^\infty \frac{(-1)^n}{n!}x^n = e^{-x} - 1. \qquad (2.30)$$

In (2.30), and throughout this book, $\mathrm{Res}\{f(z); z = z_0\}$ denotes the residue of $f(z)$ at the singularity z_0.

Application 2: One can easily verify the identity

$$\Gamma(z - n) = (-1)^n \frac{\Gamma(z)}{(1 - z)_n}, \qquad n = 0, 1, \ldots, \qquad (2.31)$$

by induction: Formula (2.31) is obviously true for $n = 0$, and the cases n and $n + 1$ can be related by the recurrence formulas (2.16) and (2.27).

Application 3: For $\alpha \neq 0$, the singularities of $x^{-z}\Gamma(\alpha z + \beta)$ are *simple* poles at $z = p_n$, where

$$p_n = -\frac{\beta + n}{\alpha}, \qquad n = 0, 1, 2, \ldots. \qquad (2.32)$$

These poles are equispaced and form a semi-infinite lattice. For the important special case of real, nonzero α, the lattice is parallel to the real axis. For any α, the corresponding residues involve *powers* of x:

$$\mathrm{Res}\{x^{-z}\Gamma(\alpha z + \beta), z = p_n\} = \frac{1}{\alpha}\frac{(-1)^n}{n!}x^{-p_n} = \frac{1}{\alpha}\frac{(-1)^n}{\Gamma(n + 1)}x^{-p_n}, \qquad n = 0, 1, 2, \ldots. \quad (2.33)$$

In the special case $\alpha > 0$ ($\alpha < 0$), the lattice continues indefinitely to the left (to the right) and the powers of x ascend (descend).

Application 4: Let us find the poles and residues of $\Gamma(z)\Gamma(z - 1)x^{-z}$.

At $z = 1$, there is a simple pole with residue $1/x$. At $z = 0, -1, -2, \ldots$, there are double poles. The residue at $z = 0$ is found by first writing

$$\Gamma(z)\Gamma(z - 1)x^{-z} = \Gamma(z)\frac{\Gamma(z)}{z - 1}x^{-z} = \frac{1}{z^2}\frac{1}{z - 1}[\Gamma(z + 1)]^2 x^{-z}. \qquad (2.34)$$

The functions $\frac{1}{z-1}$, $[\Gamma(z + 1)]^2$, and x^{-z} are analytic at $z = 0$ and can be expanded in Taylor series about that point. In particular,

$$x^{-z} = e^{-z \ln x} = 1 - z \ln x + \frac{1}{2}z^2(\ln x)^2 + O(z^3). \qquad (2.35)$$

When the aforementioned three series are multiplied, the desired residue is the coefficient of z. It is

$$\mathrm{Res}\{\Gamma(z)\Gamma(z - 1)x^{-z}; z = 0\} = \ln x + 2\gamma - 1. \qquad (2.36)$$

The residue at any other double pole can be found in a similar manner. The reader is invited to show that the final answer is

$$\text{Res}\{\Gamma(z)\Gamma(z-1)x^{-z}; z = -n\}$$

$$= \frac{x^n}{n!(n+1)!}\left[\ln x - 2\psi(n+1) - \frac{1}{n+1}\right], \qquad n = 0, 1, 2, \ldots. \qquad (2.37)$$

Besides powers of x, the residues at these *double* poles also involve the *logarithm* of x. Residue calculations like this are usually laborious. But they are important for the Mellin-transform method, and so we provide the following lemma.

Application 5 (Lemma): If $n = 0, 1, 2, \ldots$ and $g(z)$ is analytic and nonzero at $z = -n$, then $[\Gamma(z)]^2 g(z)x^{-z}$ has a double pole at $z = -n$ and the residue there is

$$\text{Res}\left\{[\Gamma(z)]^2 g(z)x^{-z}; z = -n\right\}$$
$$= \frac{x^n}{(n!)^2}\left[-g(-n)\ln x + 2\psi(n+1)g(-n) + g'(-n)\right], \qquad n = 0, 1, 2, \ldots. (2.38)$$

It is possible to derive (2.38) using the material we have already presented and the interested reader is invited to do so. A derivation that includes additional interesting information can be found in Appendix B.

Many other expressions, arising when applying the Mellin-transform method and involving gamma functions, can be written in a form appropriate for the application of the above lemma. To use the lemma to verify (2.37), for example, substitute $\Gamma(z-1)$ by $\Gamma(z)/(z-1)$. An identical substitution helps one show (2.70) from (2.68); see below. In the latter case, the algebra is facilitated by using the identity

$$\Gamma\left(\frac{3}{2} - z\right)\Gamma\left(\frac{1}{2} + z\right) = \frac{\pi(\frac{1}{2} - z)}{\cos \pi z}, \qquad (2.39)$$

which is a straightforward consequence of the recurrence and reflection formulas (2.16) and (2.18). As a last, more general example, to find the residues of $\Gamma(z-n)\Gamma(z)h(z)x^{-z}$ ($h(z)$ analytic and nonzero at $z = -n$, $n = 0, 1, 2, \ldots$), use (2.38) after substituting $\Gamma(z-n)$ by $\Gamma(z)/(z-n)_n$.

2.8   TABLE LOOKUP OF MELLIN TRANSFORMS; MELLIN–BARNES INTEGRALS

When calculating integrals with the Mellin-transform method, one needs to find the Mellin transforms of functions involved in the integrand. This is usually done using symbolic programs such as Mathematica or Matlab,* or published tables of Mellin transforms. We present our own short Table 2.1, which shows several functions $f(x)$, their Mellin transforms $\tilde{f}(z)$, and the corresponding SOA's. The reader can easily verify the SOA's using the rules of Appendix A regarding the convergence/divergence of integrals. The specific functions $f(x)$ of Table 2.1 have been selected for the simple reason that they will be useful when evaluating our example-integrals.

A primary goal of the Mellin-transform method is to find a Mellin–Barnes integral representation of the integral to be evaluated, so we proceed to discuss Mellin–Barnes integrals. What one immediately observes is that each $\tilde{f}(z)$ of Table 2.1 has been written as a product, in which the factors have the form $\Gamma(a + Az)$, $[\Gamma(a + Az)]^{-1}$, or α^{-z}, where all A's are real. For reasons to become apparent, let us call this a "standard product." The integrands of the corresponding inversion integrals will also be "standard products" multiplied by x^{-z}. For instance, as a consequence of Entry 4 of Table 2.1, the inversion formula (2.5), and property (2.6), one has

$$\left(\frac{\sin 3x}{3x}\right)^2 = -\frac{\sqrt{\pi}}{4}\frac{1}{2\pi i}\int_{\delta-i\infty}^{\delta+i\infty}\frac{\Gamma\left(\frac{z}{2}-1\right)}{\Gamma\left(\frac{3}{2}-\frac{z}{2}\right)}(3x)^{-z}\,dz, \qquad 0 < \delta < 2. \qquad (2.40)$$

Convergent integrals like the one on the right-hand side of (2.40)—with integrands of the aforementioned type, integrated along proper contours in the z-plane—are called *Mellin–Barnes integrals* [7, p. 11], [16, Section 1.19]. They are very important for us because *"most" functions $f(x)$ can be written as Mellin–Barnes integrals or as linear combinations of Mellin–Barnes integrals. In other words, their Mellin transforms $\tilde{f}(z)$ can be written as linear combinations of standard products.* It is beyond the scope of this book to make this statement precise; interested readers should consult [7].

A standard product can often be written in other "nonstandard" forms (simple illustrations of this fact are provided by the reflection and recurrence formulas (2.18) and (2.16)). Since Mellin–Barnes integrals are important for us, *it is preferable to use "standard products" when possible.* This brings us back to discussing published tables of Mellin transforms: The well-known tables by A. Erdélyi et al. [17] and F. Oberhettinger [18] do not always express their Mellin transforms as standard products (neither does the standard table of integrals [19] by I. S. Gradshteyn and I. M. Ryzhik), so it is preferable to use tables that do so. By far the most extensive such table is [6, 8.4] in the three-volume work by A. P. Prudnikov, Yu. A. Brychkov, and O. I. Marichev. Let us also mention the table in [7] as well as the much shorter tables in [13].

---

* Matlab is a registered trademark of The MathWorks, Natick, MA, USA.

TABLE 2.1: Selected Functions $f(x)$, Together With Their Mellin Transforms $\tilde{f}(z)$, and the SOAs of $\tilde{f}(z)$.

ENTRY	$f(x)$	$\tilde{f}(z)$	SOA
1	$(x^2 + 1)^{-1}$	$\frac{1}{2}\Gamma\left(\frac{z}{2}\right)\Gamma\left(1 - \frac{z}{2}\right)$	$0 < \mathrm{Re}\{z\} < 2$
2	$\begin{cases} (1 - x^2)^{-1/2}, & 0 < x < 1 \\ 0, & x > 1 \end{cases}$	$\frac{\sqrt{\pi}}{2}\dfrac{\Gamma\left(\frac{z}{2}\right)}{\Gamma\left(\frac{z}{2}+\frac{1}{2}\right)}$	$\mathrm{Re}\{z\} > 0$
3	$\begin{cases} 0, & 0 < x < 1 \\ (x^2 - 1)^{-1/2}, & x > 1 \end{cases}$	$\frac{\sqrt{\pi}}{2}\dfrac{\Gamma\left(\frac{1}{2}-\frac{z}{2}\right)}{\Gamma\left(1-\frac{z}{2}\right)}$	$\mathrm{Re}\{z\} < 1$
4	$\left(\frac{\sin x}{x}\right)^2$	$-\dfrac{\sqrt{\pi}}{4}\dfrac{\Gamma\left(\frac{z}{2}-1\right)}{\Gamma\left(\frac{3}{2}-\frac{z}{2}\right)}$	$0 < \mathrm{Re}\{z\} < 2$
5	$J_\nu(x)$	$\frac{1}{2}\left(\frac{1}{2}\right)^{-z}\dfrac{\Gamma\left(\frac{\nu}{2}+\frac{z}{2}\right)}{\Gamma\left(1+\frac{\nu}{2}-\frac{z}{2}\right)}$	$-\mathrm{Re}\{\nu\} < \mathrm{Re}\{z\} < \frac{3}{2}$
6	$J_\mu(x)J_\nu(x)$	$\frac{1}{2}\left(\frac{1}{2}\right)^{-z}\Gamma(1-z)\,\Gamma\left(\frac{\nu}{2}+\frac{\mu}{2}+\frac{z}{2}\right)\Gamma\left(1+\frac{\nu}{2}+\frac{\mu}{2}-\frac{z}{2}\right)^{-1}$ $\times\left[\Gamma\left(1+\frac{\nu}{2}-\frac{\mu}{2}-\frac{z}{2}\right)\Gamma\left(1+\frac{\nu}{2}+\frac{\mu}{2}+\frac{z}{2}\right)\right]^{-1}$ $\times\left[\Gamma\left(1-\frac{\nu}{2}+\frac{\mu}{2}-\frac{z}{2}\right)\right]^{-1}$	$-\mathrm{Re}\{\nu + \mu\} < \mathrm{Re}\{z\} < 1$

(Table adapted from Fikioris [1]).

Table 2.1 will be used shortly when we deal with our example-integrals. For the time being, the reader may wish to familiarize him/herself with the aforementioned published tables by using them to verify the entries in Table 2.1. Note the following:

(i) The table [6, 8.4] should be used in conjunction with the Mellin-transform properties of Section 2.2. For example, Entry 4 of Table 2.1 comes from [6, Entry 8.4.5.11], which is

$$\text{MT}\{\sin^n \sqrt{x}; z\} = 2^{1-n}\sqrt{\pi} \sum_{k=0}^{[(n-1)/2]} (-1)^{[n/2]+k} \binom{n}{k} \left(\frac{2}{n-2k}\right)^{2z} \frac{\Gamma(z+\gamma)}{\Gamma(1/2+\gamma-z)};$$

$$-n/2 < \text{Re}\{z\} < \gamma; \ \gamma = (1 - (-1)^n)/4, \tag{2.41}$$

and eqns. (2.8) and (2.7) (with $\alpha = 2$ and $\alpha = -2$, respectively). In (2.41), $[x]$ denotes the floor of x, i.e., the greatest integer that is less than or equal to x.

(ii) Formula [6, Entry 8.4.19.15], which can be used to derive Entry 6, can be simplified with aid of the duplication formula (2.19), resulting in one less gamma function.

CHAPTER 3

Generalized Hypergeometric Functions, Meijer *G*-Functions, and Their Numerical Computation

The Mellin-transform method often gives results in terms of the generalized hypergeometric function $_pF_q$ or the Meijer *G*-function. The most striking difference of these two "functions" with the more usual "special functions of mathematical physics" (the Bessel function J_ν, say) is that the former are much more general: $_pF_q$ and G involve many parameters and, depending on their values, often reduce to more usual special functions. For instance, J_ν can be written both as a $_pF_q$ and as a *G*-function. In fact, "very many" special functions have $_pF_q$ and/or *G*-function representations.

Reference [6] contains the two extensive tables [6, Chapt. 7] and [6, 8.4.52] which can be searched in a systematic manner to see whether a given $_pF_q$ or *G*-function can be reduced to a more usual special function. They are obviously very useful for our purposes and will be referred to as *reduction tables* for the $_pF_q$ and G, respectively. Extensive "reduction tables" can also be found online; see [20]. We now proceed to discuss $_pF_q$ and G in more detail.

3.1 DEFINITIONS

The *generalized hypergeometric series* of *order* (p, q) is defined as a power series in z and is denoted by $_pF_q(\alpha_1, \alpha_2, \ldots, \alpha_p; \beta_1, \beta_2, \ldots, \beta_q; z)$. The expressions for the power-series coefficients involve the p numbers α_l and the q numbers β_l ($p, q = 0, 1, \ldots$), called *upper* and *lower* *parameters*, respectively. The precise definition is

$$_pF_q(\alpha_1, \ldots, \alpha_p; \beta_1, \ldots, \beta_q; z) = \sum_{n=0}^{\infty} \frac{(\alpha_1)_n (\alpha_2)_n \cdots (\alpha_p)_n}{(\beta_1)_n (\beta_2)_n \cdots (\beta_q)_n} \frac{z^n}{n!}, \qquad (3.1)$$

where all lower parameters are assumed different from $0, -1, -2, \ldots$.* In (3.1), and throughout this book, empty products or sums are to be interpreted, in the usual manner, as unity or zero,

---

* This restriction is required because $(-m)_n$ vanishes for sufficiently large n; to understand why, take the limit $z \to m + 1$ ($m = 0, 1, \ldots$) in (2.31).

respectively. Note that the order is reduced when an upper parameter is equal to a lower parameter.

For the defining series to make sense, it must converge, at least for some z. Application of the ratio test for power series and Stirling's formula yields the following cases [6, 7.2.3].

Case F1: When $p \leq q$, the series converges for all z. In other words, the radius of convergence is infinite. This series defines, for all complex values of z, the so-called *generalized hypergeometric function*.

Case F2: When $p = q + 1$, the series converges inside the unit z-circle and diverges outside (for the behavior on the boundary $|z| = 1$, see [6, 7.2.3]) so that the radius of convergence here equals 1. In this case, the *generalized hypergeometric function*—denoted, again, by $_pF_q(\alpha_1, \ldots, \alpha_p; \beta_1, \ldots, \beta_q; z)$—is defined (i) by the series (3.1) when $|z| < 1$; (ii) by the analytic continuation of the series (3.1) when $|z| \geq 1$. The function thus defined is often (but not always) multivalued, with a branch point at $z = 1$; in those cases, the symbol $_pF_q(\alpha_1, \ldots, \alpha_p; \beta_1, \ldots, \beta_q; z)$ denotes the principal branch, as defined in [6, 7.2.3].

We finally give a condition for divergence: When $p \geq q + 2$, the series diverges for all nonzero z (zero radius of convergence.)

The *Meijer G-function* is a special type of Mellin–Barnes integral in which all coefficients A of the factors $\Gamma(a + Az)$ and $[\Gamma(a + Az)]^{-1}$ are 1 or -1. A definition adequate for the purposes of this book (adapted from [21, Section 2.1]) is

$$
G^{mn}_{pq} \left(x \left| \begin{matrix} \alpha_1, \ldots, \alpha_p \\ \beta_1, \ldots, \beta_q \end{matrix} \right. \right) = \frac{1}{2\pi i} \int_L \frac{\Gamma(\beta_1 + z) \cdots \Gamma(\beta_m + z)}{\Gamma(\alpha_{n+1} + z) \cdots \Gamma(\alpha_p + z)}
$$

$$
\times \frac{\Gamma(1 - \alpha_1 - z) \cdots \Gamma(1 - \alpha_n - z)}{\Gamma(1 - \beta_{m+1} - z) \cdots \Gamma(1 - \beta_q - z)} x^{-z} \, dz, \qquad (3.2)
$$

where m, n, p, q are integers with $0 \leq m \leq q$ and $0 \leq n \leq p$. It is assumed in this definition that the poles of $\Gamma(\beta_l + z)$, $l = 1, \ldots, m$ (let us call these "left poles"), are separated by the poles of $\Gamma(1 - \alpha_l - z)$, $l = 1, \ldots, n$ ("right poles"), by a vertical strip, as illustrated in Fig. 3.1. We note that more general definitions [6, 8.2] do not require such a separation of the poles.

As illustrated in Fig. 3.1, the path of integration in (3.2) is one of the following three types: (i) $L = L_V$, where L_V is vertical and lies within the aforementioned vertical strip. (ii) $L = L_L$, where L_L is a counterclockwise oriented loop that begins at $-\infty$, encircles all left poles but no right poles, and ends at $-\infty$ again. (iii) $L = L_R$, where L_R is a clockwise oriented loop that begins at $+\infty$, encircles all right poles but no left poles, and ends at $+\infty$ again.

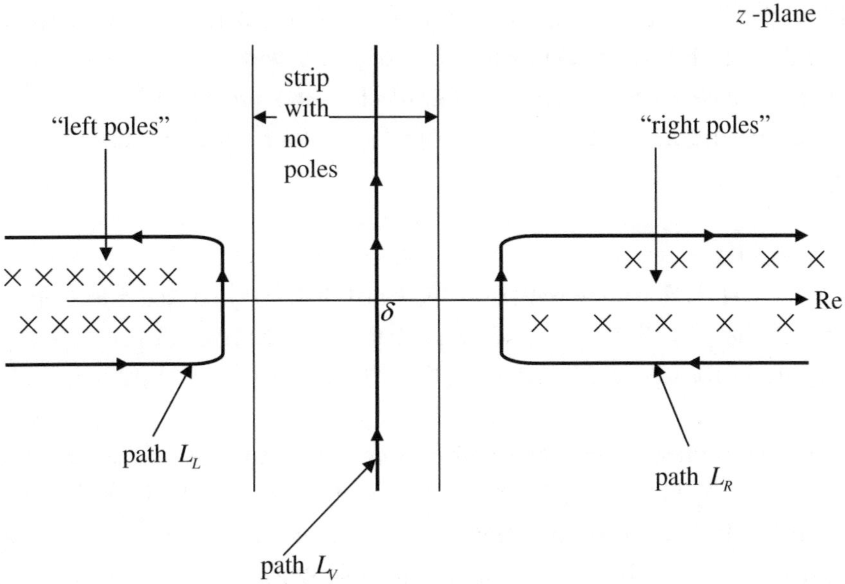

FIGURE 3.1: In definition (3.2) of G, the poles of $\Gamma(\beta_l + z)$ ("left poles," i.e., the semi-infinite pole lattices continuing indefinitely to the left) are separated by the poles of $\Gamma(1 - \alpha_l + z)$ ("right poles") by a vertical strip, within which lies the vertical path L_V; that path extends from $\delta - i\infty$ to $\delta + i\infty$. The path $L_L(L_R)$ is a loop enclosing all left poles (all right poles), but no right poles (no left poles)

We give three sets of conditions for the integral defined by (3.2) to be convergent. Define the parameters

$$B = m + n - \frac{1}{2}(p + q), \qquad C = \sum_{l=1}^{q} \beta_l - \sum_{l=1}^{p} \alpha_l. \tag{3.3}$$

The integral *converges* in the following cases [21, Section 2.1]:

Case G1: $L = L_V$, $|\arg x| < B\pi$, $B > 0$.

Case G2: $L = L_V$, $|\arg x| = B\pi$, $B \geq 0$, $p = q$, $\mathrm{Re}\{C\} < -1$.

Case G3: $L = L_V$, $|\arg x| = B\pi$, $B \geq 0$, $p \neq q$, $(p - q)\delta > \mathrm{Re}\{C\} + 1 - \frac{1}{2}(q - p)$. Here, as shown in Fig. 3.1, L_V starts at $\delta - i\infty$ and ends at $\delta + i\infty$.

Case G4: $L = L_L$, x arbitrary, $q \geq 1$, $q > p$.

Case G5: $L = L_L$, $|x| < 1$, $p = q$.

Case G6: $L = L_R$, x arbitrary, $p \geq 1$, $p > q$.

Case G7: $L = L_R$, $|x| > 1$, $p = q$.

These sufficient conditions for convergence are shown in [21, Section 2.1] by a systematic study of the integrand on the basis of Stirling's formula. *It often happens that more than one path can be used; in such cases, the G-functions defined by the different paths are one and the same* [21, Section 2.1]. Note that transforming an L_V-contour to an L_L-contour is what we have termed "closing the contour at left."

3.2   REMARKS

Every $_pF_q$ has a G-function representation [6, Entry 8.4.51.1] so that the G-function is a generalization of the $_pF_q$. Both $_pF_q$ and G possess a vast number of properties. The most extensive lists are in [6, Chaps. 7 and 8] and online in [20]; these lists include "reduction tables."

One often encounters series in theoretical work. *It is always beneficial to attempt to identify a given series with a $_pF_q$* because of the many tabulated properties of $_pF_q$ and also (as will be discussed immediately) because of the convenience in numerical evaluation of the $_pF_q$.

It is usual to deal with Mellin–Barnes integrals whose coefficients A of the factors $\Gamma(a + Az)$ and $[\Gamma(a + Az)]^{-1}$ are all *rational* numbers. As a first step in writing these as G-functions, change the integration variable to yield *integer* coefficients A and then use the multiplication formula (20).

3.3   NUMERICAL COMPUTATION OF $_pF_q$ AND G

There now exist packaged routines for the numerical calculation of $_pF_q$ and G. Such routines should greatly enhance the use of $_pF_q$ and G in engineering applications. For instance, Mathematica 5.0 can numerically compute both $_pF_q$ and G. Matlab 7.0 can handle $_pF_q$, but not G. For numerical computation, today's packaged routines do not rely exclusively on the definitions, but rather on the numerous properties mentioned above. To quickly persuade oneself of this, note that both Mathematica and Matlab can handle $_pF_q$ when $|z| > 1$ in Case F2. *When numerical results are of primary concern, it is today often sufficient to express the quantity of interest in terms of $_pF_q$ or G and to use the aforementioned routines as black boxes.*

We close this section by noting that there are further generalizations of the G-function. Such is the Fox H-function, in which the coefficients A in $\Gamma(a + Az)$ and $[\Gamma(a + Az)]^{-1}$ need not be 1 or -1. This function would be very useful for the Mellin-transform method but, to the best of the author's knowledge, it cannot be calculated by today's packaged routines. For brief discussions of H, see [13] and [21]; for more comprehensive expositions, see [6, 8.3].

CHAPTER 4

The Mellin-Transform Method
of Evaluating Integrals

4.1 A GENERAL DESCRIPTION OF THE MELLIN-TRANSFORM METHOD

We finally come to the Mellin-transform method itself. It applies to integrals $f(x)$ that are Mellin convolutions. That is, the integral $f(x)$ to be evaluated can be written as

$$f(x) = \int_0^\infty g(xy)h(y)\, dy = (g \oslash h)(x), \qquad x > 0, \qquad (4.1)$$

where x is a positive parameter. Many integrals have this form (all Laplace transforms and all Fourier cosine transforms, for example), or can easily be written in terms of integrals having this form (all Fourier transforms, for example). We first give a general (but sketchy) description of the Mellin-transform method.

Initial step: Apply formula (2.11) to obtain a complex-integral representation of $f(x)$:

$$f(x) = \frac{1}{2\pi i} \int_{\delta-i\infty}^{\delta+i\infty} \tilde{g}(z)\tilde{h}(1-z)x^{-z}\, dz, \qquad (4.2)$$

which will, hopefully, be an Mellin–Barnes integral. By the discussion in Section 2.3, this step is the same as taking the Mellin transform of (4.1) (with respect to x), using formula (2.13) and, finally, the inversion formula (2.5), in which $\tilde{f}(z) = \tilde{g}(z)\tilde{h}(1-z)$. In (4.2), δ belongs both to the SOA of $\tilde{g}(z)$ *and* to the SOA of $\tilde{h}(1-z)$, which must overlap.

Other steps: The aforementioned Mellin–Barnes integral representation of $f(x)$ is the heart of the Mellin-transform method, and having obtained it, one can proceed in many ways. First of all, a representation of $f(x)$ as a G-function is often obtainable from (4.2) with very little effort. Second, (4.2) can often yield a series expansion: In (4.2), determine the singularities to the *left* of the contour, which will hopefully be poles (but not necessarily *simple* poles) located at $z = z_0, z_1, \ldots$. Then, close the contour at left (we will postpone discussing conditions under which this is possible until Chapter 8) and apply the residue theorem to obtain

$$f(x) = \sum_{n=0}^{\infty} \text{Res}\{\tilde{f}(z)x^{-z}; z = z_n\}. \tag{4.3}$$

Typically, such series are *ascending* series expansions; often (but not always, as we will see) one can identify the series with a $_pF_q$. We now present a first example illustrating the Mellin-transform method.

4.2   A FIRST EXAMPLE

Our first example is the integral

$$f(x) = \int_0^1 \left(\frac{\sin xy}{xy}\right)^2 \frac{1}{\sqrt{1-y^2}} \, dy, \tag{4.4}$$

which comes up when calculating the conductance of a two-dimensional slot with constant aperture field [22, p. 720], [23]. For more on the origins of this integral, see Chapter 6.

Let us, in this first example, apply the Mellin-transform method without omitting details. The integral $f(x)$ can be written as in (4.1), where

$$g(x) = \left(\frac{\sin x}{x}\right)^2 \tag{4.5}$$

and

$$h(x) = \begin{cases} (1 - x^2)^{-1/2}, & \text{if } 0 < x < 1, \\ 0, & \text{if } x > 1. \end{cases} \tag{4.6}$$

The Mellin transform $\tilde{g}(z)$ can be found directly as Entry 4 of Table 4.1. From Entry 2, we deduce that

$$\tilde{h}(1 - z) = \frac{\sqrt{\pi}}{2} \frac{\Gamma(\frac{1}{2} - \frac{z}{2})}{\Gamma(1 - \frac{z}{2})}, \qquad \text{Re}\{z\} < 1. \tag{4.7}$$

The SOAs of $\tilde{g}(z)$ and $\tilde{h}(1 - z)$ do overlap; the strip of overlap is $0 < \text{Re}\{z\} < 1$ so that (4.2) gives

$$f(x) = -\frac{\pi}{8} \frac{1}{2\pi i} \int_{\delta-i\infty}^{\delta+i\infty} \frac{\Gamma(\frac{z}{2} - 1)\Gamma(\frac{1}{2} - \frac{z}{2})}{\Gamma(1 - \frac{z}{2})\Gamma(\frac{3}{2} - \frac{z}{2})} x^{-z} \, dz, \qquad 0 < \delta < 1, \tag{4.8}$$

which is a Mellin–Barnes integral representation of $f(x)$.

Each gamma function in (4.8) contributes to the integrand a semi-infinite lattice of poles (if the function is in the numerator) or zeros (if in the denominator). The locations of these poles can be determined using Application 3 of Section 2.7 and the results are summarized in

TABLE 4.1: Pole/Zero Contributions to Integrand of (4.8) (Section 4.2)

GAMMA FUNCTION	N OR D	−6	−5	−4	−3	−2	−1	0	1	2	3	4	5	6
$\Gamma\left(\frac{z}{2}-1\right)$	N	...		P		P		P	P					
$\Gamma\left(\frac{1}{2}-\frac{z}{2}\right)$	N								P	P	...			
$\Gamma\left(1-\frac{z}{2}\right)$	D									Z	Z	...		
$\Gamma\left(\frac{3}{2}-\frac{z}{2}\right)$	D									Z	...			

If a gamma function is in the numerator N (or denominator D), it contributes a pole P (or zero Z) at the location specified. Ellipses (...) at right (or left) indicates that a particular Pole/Zero lattice continues indefinitely to the right (or left). The thick line between 0 and 1 indicates that the inversion path lies in the strip $0 < \operatorname{Re}\{z\} < 1$.

Table 4.1. Evidently, a pole contribution at a specified location cancels a zero contribution at the same location—for example, there is no pole or zero at $z = 3$. The final conclusion from Table 4.1 is that, to the *left* of the inversion path, there are *simple* poles at $0, -2, -4, \ldots$. For reasons to be discussed in Chapter 8, it is possible to close the contour at left. Therefore, (4.3) is, in our case,

$$f(x) = -\frac{\pi}{8} \sum_{n=0}^{\infty} \operatorname{Res}\left\{\frac{\Gamma(\frac{z}{2}-1)\Gamma(\frac{1}{2}-\frac{z}{2})}{\Gamma(1-\frac{z}{2})\Gamma(\frac{3}{2}-\frac{z}{2})} x^{-z}; z = -2n\right\}. \qquad (4.9)$$

To evaluate the residues, set $z = -2n$ except in $\Gamma(\frac{z}{2}-1)$ to which the poles at $z = -2n = p_{n+1}$, $(n = 0, 1, \ldots)$ are due. Then, use Application 3 of Section 2.7 once again:

$$f(x) = -\frac{\pi}{8} \sum_{n=0}^{\infty} \frac{\Gamma(\frac{1}{2}+\frac{2n}{2})}{\Gamma(1+\frac{2n}{2})\Gamma(\frac{3}{2}+\frac{2n}{2})} \operatorname{Res}\left\{\Gamma\left(\frac{z}{2}-1\right) x^{-z}; z = p_{n+1}\right\}$$

$$= \frac{\pi}{4} \sum_{n=0}^{\infty} \frac{\Gamma(\frac{1}{2}+n)}{\Gamma(1+n)\Gamma(\frac{3}{2}+n)} \frac{(-1)^n}{\Gamma(n+2)} x^{2n}. \qquad (4.10)$$

Now set $\Gamma(1 + n) = n!$ and express the three remaining gamma functions in terms of Pochhammer's symbol using (2.26). By Section 2.4, $\Gamma(1) = 1$ and $\Gamma(\frac{1}{2}) = 2\Gamma(\frac{3}{2}) = \sqrt{\pi}$. Finally, compare with the definition (3.1) of $_pF_q$ (Case F1 of Section 3.1) to obtain

$$f(x) = \frac{\pi}{2} {}_1F_2\left(\frac{1}{2}; \frac{3}{2}, 2; -x^2\right). \qquad (4.11)$$

Equation (4.11) (or the equivalent series form (4.10) which, as discussed in Section 3.2, is less preferable for numerical calculation by modern routines) is our final result. Let us note that Entry 7.14.2.46 of the "reduction table" in [6, Chapt. 7] actually provides a "simplified" answer. But that answer involves rather unusual special functions (a Laguerre function and a modified Struve function), so it will not be repeated here.

As discussed in Appendix C, the series form (4.10) can also be determined by more direct methods. Nonetheless, neither [22] nor [23] contain an evaluated expression for $f(x)$. We have thus completed our first example, which was completely straightforward. An equally simple example (it is an integral arising in a rain attenuation problem [24]) which the reader may wish to work on his/her own is provided in Section D.1 of Appendix D.

CHAPTER 5

Power Radiated by Certain Circular Antennas

This and the next two chapters present further antenna/electromagnetics problems to which the Mellin-transform method can be applied. In each chapter, we give background information, use our method to evaluate the associated integral exactly, and discuss, apply, or interpret the exact results. Most of the exact results that follow (as well as those that precede) have been verified numerically. That is, the final result agrees with numerical evaluation of the original integral. Such checks are always a good idea when possible. Relations of our exact results to integrals tabulated in the standard tables [4]–[6] and [19] are mentioned in Section 9.5 of our discussion in Chapter 9, while alternative derivations of some of our exact results are provided in Appendix C.

5.1 CONSTANT-CURRENT CIRCULAR-LOOP ANTENNAS

Circular, thin-wire loop antennas are one of the most basic types of radiators and are discussed in standard textbooks, e.g., [22, chapt. 5]. Simple in construction, they are used for frequencies from about 3 MHz up to microwave. Electrically small loops are rather poor radiators (the radiation resistance R_r is usually smaller than the loss resistance) and are therefore used when efficiency is not of primary importance. Large loops have a larger R_r (our result (5.10) will illuminate this) and are therefore used primarily as elements of directional arrays—such as helical antennas and Yagi–Uda arrays—with the loop circumference and interelement spacing chosen to achieve the desired directional properties.

Many studies have focused on the case of *constant* loop current I_0. Such studies are practically useful for at least two reasons: For sufficiently small loops (and, also, for large inter-element spacings in the case where the loop is an array element) the current is truly constant. Second, constant current distributions can be achieved even for large loops [22, p. 249]: one divides the loop into sections and feeds each section with a different feed line. Often, all lines are driven from a common source.

An accurate method for determining the field radiated by a constant-current, circular-loop, thin-wire antenna proceeds from the standard integral [22, eqn. (5–14)] for the vector potential **A**, which is ϕ-directed: The distance from loop to observation point is approximated [22, eqn. (5–43)] subject to the usual condition $r \gg a$, where a is the loop radius and (r, θ, ϕ) are spherical coordinates with origin at the loop's center and z-axis perpendicular to the loop. This leads to an integral which can be evaluated in terms of the Bessel function J_1. The resulting expression is then used in the familiar formulas [22, Section 3.6] relating **A** to the radiated fields. With $\zeta_0 = 120\pi \ \Omega$, the nonzero components are [22, eqn. (5–54)]

$$E_\phi = -\zeta_0 H_\theta = \frac{I_0 ka \ \zeta_0 e^{-jkr}}{2r} J_1(ka \sin \theta). \qquad (5.1)$$

The radiated power [22, eqn. (5–58)], found by integrating over a large sphere, therefore equals $\pi (ka)^2 \zeta_0 |I_0|^2 f(ka, 1, 1, 1)/2$, with the more general integral $f(x, \mu, \nu, \tau)$ defined by

$$f(x, \mu, \nu, \tau) = \int_0^{\pi/2} J_\mu(x \sin \theta) J_\nu(x \sin \theta) \sin^\tau \theta \ d\theta, \qquad x > 0. \qquad (5.2)$$

The reason for the more general notation will be explained in the next section. Once the radiated power is found, the directivity and radiation resistance R_r easily follow [22, Section 5.3.2].

5.2   CIRCULAR-PATCH MICROSTRIP ANTENNAS; CAVITY MODEL

The cavity model is one of the most popular methods for the analysis of circular microstrip antennas [22, ch. 14]. One treats the region between patch and ground plane as a cavity bounded above and below by electric conductors and by a magnetic conductor along the patch's perimeter. Within this model, the radiated power can be shown to be proportional to the quantity [22, eqns. (14–75), (14–72)]

$$2f(ka, 0, 0, 1) - f(ka, 0, 0, 3) - 2f(ka, 0, 2, 3) + 2f(ka, 2, 2, 1) - f(ka, 2, 2, 3), \qquad (5.3)$$

where $f(x, \mu, \nu, \tau)$ was defined in (5.2) and where a is the "effective radius" [22, eqn. (14–67)] of the patch. Details of the derivation of (5.3) are provided in [22, Section 14.3], [25], and [26]. With the radiated power determined, one can immediately find the directivity; see [22, eqn. (14–80)].

The integral f in (5.2) thus comes up in at least two antenna problems. f has deserved a great deal of attention: Recently, f has been the subject of much discussion in the *IEEE Antennas and Propagation Magazine* [27]–[34]. Some of these papers are referred to in the standard textbook [22]. For the loop antenna ($\mu = \nu = \tau = 1$), Balanis [22] proposes numerical evaluation of (5.2) and provides a computer program for doing so. Reference [30] mentions an additional application in which f arises, namely, the circular loop with a cosinusoidal current. Finally, the exact evaluation and/or the asymptotics of f (more precisely, of special or of more general cases of f) have been much discussed in other (more mathematical) contexts [35]–[41]. In the next section, we provide a closed-form expression for f, in terms of a $_pF_q$, by straightforward application of the Mellin-transform method.

5.3   INTEGRAL EVALUATION

Change the variable $\sin\theta = y$ in (5.2) to obtain the more suitable expression

$$f(x, \mu, \nu, \tau) = \int_0^1 \frac{J_\mu(xy)J_\nu(xy)}{\sqrt{1-y^2}} y^\tau \, dy, \qquad x > 0, \qquad (5.4)$$

which is (4.1) with $g(x) = J_\mu(x)J_\nu(x)$, $h(x) = x^\tau(1-x^2)^{-1/2}$ for $0 < x < 1$, and $h(x) = 0$ for $x > 1$. To avoid unnecessary complications, let us assume that the complex quantities μ, ν, and τ satisfy

$$\text{Re}\{\mu\} > 0, \quad \text{Re}\{\nu\} > 0, \quad \text{Re}\{\tau\} > 0. \qquad (5.5)$$

By the rules of Appendix A, the restrictions in (5.5) are sufficient for the integral in (5.4) to convergence.

Because of the Bessel functions, the integral in (5.4) might appear more difficult than our very first integral, eqn. (4.4). With the Mellin transforms $\tilde{g}(z)$ and $\tilde{h}(z)$ obtainable from Table 2.1 and eqn. (2.7), however, the Mellin-transform method can be applied just as before. We omit lengthy intermediate formulas and directly give the result as a Mellin–Barnes integral

$$f(x, \mu, \nu, \tau) = \frac{\sqrt{\pi}}{4} \frac{1}{2\pi i} \int_{\delta-i\infty}^{\delta+i\infty} \frac{\Gamma(1-z)\Gamma(\frac{\nu}{2}+\frac{\mu}{2}+\frac{z}{2})}{\Gamma(1+\frac{\nu}{2}-\frac{\mu}{2}-\frac{z}{2})\Gamma(1+\frac{\nu}{2}+\frac{\mu}{2}-\frac{z}{2})}$$

$$\times \frac{\Gamma(\frac{1}{2}-\frac{z}{2}+\frac{\tau}{2})}{\Gamma(1-\frac{\nu}{2}+\frac{\mu}{2}-\frac{z}{2})\Gamma(1-\frac{z}{2}+\frac{\tau}{2})} \left(\frac{x}{2}\right)^{-z} dz,$$

$$-\text{Re}\{\nu+\mu\} < \delta < 1. \qquad (5.6)$$

Once again, we have *simple* poles to the left of the contour, contributed here by $\Gamma(\frac{\nu}{2}+\frac{\mu}{2}+\frac{z}{2})$. Closing the contour at left (see Chapter 8) and calculating residues leads to a convergent series. With the duplication formula (2.19) and the definition (3.1) for the $_pF_q$, the series can be identified with a $_3F_4$. The result is

$$f(x, \mu, \nu, \tau) = \frac{\sqrt{\pi}}{2} \left(\frac{x}{2}\right)^{\mu+\nu} \frac{\Gamma(\lambda)}{\Gamma(\mu + 1)\Gamma(\nu + 1)\Gamma(\lambda + \frac{1}{2})}$$

$$\times {}_3F_4\left(\frac{1}{2} + \frac{\nu}{2} + \frac{\mu}{2}, 1 + \frac{\nu}{2} + \frac{\mu}{2}, \lambda; \mu + 1, \nu + 1, \mu + \nu + 1, \lambda + \frac{1}{2}; -x^2\right),$$

$$(5.7)$$

where

$$\lambda = \frac{1 + \nu + \mu + \tau}{2}. \qquad (5.8)$$

For general μ, ν, and τ in (5.7), the "reduction table" in [6, Chapt. 7] gives no simpler form, so that (5.7) is our final result.

The series form corresponding to (5.7) can be determined by more direct methods, as described in [30] and Appendix C. Nonetheless, neither [22] nor [27]–[34] mention the $_3F_4$ whose use—as discussed in Chapter 3 and, further, in the next Section 5.4—presents several advantages.

For many of the cases in [22] and [27]–[34], it is possible to lower the order in (5.7). For instance, the $_3F_4$ reduces to a $_2F_3$ when $\nu = \mu$ and further reduces to a $_1F_2$ when, also, $\tau = 1$:

$$f(x, \mu, \mu, 1) = \frac{x^{2\mu}}{\Gamma(2\mu + 2)} {}_1F_2\left(\mu + \frac{1}{2}; 2\mu + 1, \mu + \frac{3}{2}; -x^2\right). \qquad (5.9)$$

The reduction table in [6, Chapt. 7] gives certain simpler forms for special cases of (5.9)—especially when $\mu = 0$ or $\mu = 1$—but, once again, those forms involve rather unusual special functions and will not be repeated here.

The fact that the two Bessel functions in (5.4) have identical arguments is essential for the success of the Mellin-transform method (or, indeed, for the alternative procedure of Appendix C); for related discussions (but for a different integral), see [42, Section 6.1].

5.4   APPLICATION TO ELECTRICALLY LARGE LOOP ANTENNAS

We return to the loop antenna with $ka = C/\lambda$, where C is the circumference, so that the relevant integral f equals the expression in (5.9) with $\mu = 1$. The first few terms in the definition (3.1) for the $_1F_2$ easily provide a small-C/λ approximation for the power or for R_r. We do not dwell on this. Instead, we focus on the nontrivial case where C/λ is *large*. We use two terms of the large-argument asymptotic expansion of the $_1F_2$, which can be found in [43]. The following

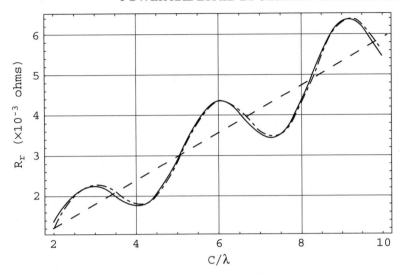

FIGURE 5.1: Radiation resistance R_r of circular loop as function of circumference C/λ: Exact R_r (solid line), together with linear approximation (i.e., first term of (5.10); dashed curve) and full approximation (5.10) (dot-dashed curve). *(Figure adapted from Fikioris [1])*

asymptotic formula for the radiation resistance R_r is thus easily obtained

$$R_r \sim 60\pi^2 \frac{C}{\lambda}\left[1 + \frac{1}{\sqrt{\pi}}\left(\frac{C}{\lambda}\right)^{-1/2}\cos\left(2\frac{C}{\lambda} + \frac{\pi}{4}\right)\right]\Omega . \qquad (5.10)$$

The first (linear) term $60\pi^2(C/\lambda)$ is the approximation provided as [22, eqn. (5–63a)] (derived in [22] directly from the integral). The second term grows and oscillates. Figure 5.1 shows the exact R_r (as calculated from (5.9)—numerical integration of (5.4) of course gives a coincident curve), together with the first (linear) term, and the full approximation (5.10). It is seen that the previously published approximation is greatly improved (compare also to [22, Fig. 5.10]). In fact, (5.10) sheds light on the interesting way in which R_r grows, and provides very good quantitative accuracy: The error is less than 5% even for C/λ as small as 4, and the error decreases (nonmonotonically) as C/λ increases.

CHAPTER 6

Aperture Admittance of a 2-D Slot Antenna

Aperture antennas, especially rectangular ones, are very common at microwave frequencies. Many analysis methods [22, Chapt. 12] assume an infinite, planar, perfectly conducting ground plane with a known tangential aperture field E_{tan}, and proceed to determine the complete fields from Maxwell's equations. One often assumes that E_{tan} is constant and parallel to the rectangle's small side [22, Sections 12.5 and 12.9]. For simplicity, it is sometimes further assumed that the rectangle is infinitely long [22, p. 718], [23], as in a parallel-plate waveguide with 90-degree bends; the radiated fields in this simpler 2-D problem approximate those of long, narrow rectangular slots. Let our 2-D slot lie on the xy-plane, with width b parallel to the y-axis and with $E_{\text{tan}} = E_y = E_0$. This assumed field corresponds to the dominant (TEM) field in an infinitely long, parallel-plate waveguide.

The complete fields are most easily found by the spectral-domain method, which in this case amounts to taking a Fourier transform in y. If $\mathcal{E}_y(k_y)$ and $\mathcal{H}_x(k_y)$ are the transforms of the tangential, on-aperture, spatial-domain fields $E_y(y) = E_0$ and $H_x(y)$, the transforms turn out to be [22, p. 718]

$$\mathcal{E}_y(k_y) = \frac{\zeta_0}{k}\sqrt{k^2 - k_y^2}\,\mathcal{H}_x(k_y) = b\,E_0\frac{\sin(k_y b/2)}{(k_y b/2)}. \tag{6.1}$$

The aperture admittance $Y_a = G_a + jB_a$ is defined by adapting the equation $Y_a = 2P^*/|V|^2$ from ordinary circuit theory: Here, $V = b\,E_0$ is the aperture voltage and P is the radiated power per unit length, determined by integrating $E_y H_x = E_0 H_x$ along y. By Parseval's theorem, P can also be found from the spectral-domain fields as $P = \int_{-\infty}^{\infty} \mathcal{E}_y \mathcal{H}_x^* \, dk_y$. Substituting (6.1) and taking the *imaginary part* shows that the susceptance B_a is $B_a = 2f(kb/2)/(\lambda\zeta_0)$, where f is the integral [22, p. 720], [23]:

$$f(x) = \int_1^{\infty} \left(\frac{\sin xy}{xy}\right)^2 \frac{1}{\sqrt{y^2 - 1}}\,dy, \qquad x > 0. \tag{6.2}$$

We note that the *real part*—which, when multiplied by $2/(\lambda \zeta_0)$ equals the conductance G_a [22, p. 720], [23]—is our very first example-integral (Section 4.2, eqn. (4.4), with $x = kb/2$). Neither [22] nor [23] contain an evaluated form for (6.2) (or, as we already mentioned, for (4.4)).

The integral in (6.2) is more interesting than those of eqns. (4.4) and (5.4) in two respects: (i) Because of the infinite integration interval and the slowly-decaying, oscillatory integrand, direct numerical evaluation of (6.2) is less straightforward (i.e., it is less accurate and requires more computer time, as further discussed in Section 9.6) and (ii) it is much more difficult to come up with our final result (eqn. (6.5) below) using other methods.

With the aid of (4.1), (4.2), and Table 2.1, the expression as a Mellin–Barnes integral turns out to be

$$f(x) = -\frac{\pi}{4} \frac{1}{2\pi i} \int_{\delta-i\infty}^{\delta+i\infty} \frac{\Gamma(z-1)\Gamma(z)}{\Gamma(\frac{3}{2}-z)\Gamma(\frac{1}{2}+z)} x^{-2z} dz, \qquad 0 < \delta < 1, \qquad (6.3)$$

in which a change of variable was made so that the coefficients of z in the gamma functions are 1 or -1. To identify with a G-function, indent the contour to the right to pick up the residue from $z = 1$. Determine that residue using Application 3 of Section 2.7. Close the contour at left and compare with (3.2) and Case G4 of Section 3.1 to obtain

$$f(x) = \frac{1}{2x^2} - \frac{\pi}{4} G_{13}^{20}\left(x^2 \left|\begin{matrix} \frac{1}{2} \\ 0 & -1 & -\frac{1}{2} \end{matrix}\right.\right). \qquad (6.4)$$

There is no simplification of (6.4) in the "reduction table" [6, 8.4.52]. As discussed in Section 3.3, it is possible to use (6.4), directly, for numerical computation. But we can also proceed from (6.3) to find a more classical—and in a sense more revealing—expression than (6.4) as follows.

As discussed in Chapter 8, one can close the contour of (6.3) at left. Within the closed contour, there are *double* poles located at $z = 0, -1, -2 \dots$. The residues at these poles can be found as in Application 4 of Section 2.7 or, more systematically, with the aid of the lemma (Application 5) of Section 2.7. One thus obtains

$$f(x) = \sum_{n=0}^{\infty} \frac{(-x^2)^n}{(n!)^2(n+1)(2n+1)} \left[-\ln x + \psi(n+1) + \frac{4n+3}{2(n+1)(2n+1)} \right]. \qquad (6.5)$$

The ascending series of (6.5) involves *two* convergent power series, one of which is multiplied by $\ln x$. We stress that *the logarithm appears because of the double poles in the integrand of the Mellin–Barnes integral.* Note that the series multiplying $\ln x$ can be identified with a ${}_pF_q$ (coincidentally, it is the same ${}_pF_q$ that occurs in (4.11)), but not the other series.

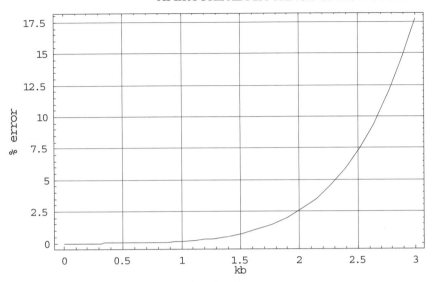

FIGURE 6.1: Percentage error between the approximate result (6.6) and the exact result, as function of slot width kb

For narrow slots (that is, small values of $x = kb/2$), the series in (6.5) converges very rapidly and is particularly useful for numerical computation (at least in this example; there is no beforehand guarantee that a series arising from the Mellin-transform method will converge rapidly.) To illustrate, when $kb = 10$, one gets an accuracy of 3% with 12 terms, 0.005% with 15 terms, and 0.0005% with 16 terms. When kb is smaller, fewer terms are required: With two terms, the approximation for the aperture susceptance B_a is

$$B_a \sim \frac{2}{\lambda \zeta_0} \left[-\ln \frac{kb}{2} - \gamma + \frac{3}{2} - \frac{(kb)^2}{24} \left(-\ln \frac{kb}{2} - \gamma + \frac{19}{12} \right) \right]. \tag{6.6}$$

Formula (6.6), which is simple enough for "back-of-the-envelope" calculations, is an improvement to the "quasi-static result" of [23], [22, p. 720]. The quasi-static result essentially corresponds to keeping one term in (6.5). The improvement is significant: With $kb = 2$, there is a 2.6% error with two terms compared to a 19% error with one term. When $0 < kb < 3$, the error is seen in Fig. 6.1 to decrease monotonically and rapidly as kb decreases.

CHAPTER 7

An Integral Arising in the Theory of Biaxially Anisotropic Media

The unpublished studies [44], [45] (which are somewhat similar to the recent papers [46], [47]) deal with the Green's function in unbounded, biaxially anisotropic media, with the aim of understanding the behavior of the two types of waves possible in such media. The problem is interesting in that, in its usual form [48], the Sommerfeld radiation condition is not applicable; that form requires isotropic media. To replace the radiation condition, Toumpis *et al.* [44]–[47] (see also [49]) initially assume a small loss, choose the solution that is bounded at infinity, and, finally, take the limit of that solution for zero loss.

In [44], the Green's function is known in cylindrical coordinates through its inverse Fourier transform. The integral over the radial Fourier variable is then performed. A key constituent of the resulting expression is the integral [44], [45]

$$f_{2m+1}(x) = \int_0^\infty \frac{J_{2m+1}(xy)}{y^2 + 1} \, dy, \quad x > 0, \quad m = 0, 1, 2, \ldots, \tag{7.1}$$

which we will evaluate using the Mellin-transform method combined with additional manipulations. Our treatment here is more advanced, but the final answer (7.12) will be particularly simple.

For reasons to become apparent, we will first deal with the more general integral obtained by replacing the odd, positive integer $2m + 1$ by a complex parameter v, viz.,

$$f_v(x) = \int_0^\infty \frac{J_v(xy)}{y^2 + 1} \, dy, \quad x > 0, \quad \mathrm{Re}\{v\} > -1, \tag{7.2}$$

and take the limit $v \to 2m + 1$ as a final step. With the aid of Table 2.1, one can come up with the Mellin–Barnes integral expression

$$f_v(x) = \frac{1}{4} \frac{1}{2\pi i} \int_{\delta - i\infty}^{\delta + i\infty} \frac{\Gamma(\frac{v}{2} + \frac{z}{2})\Gamma(\frac{1}{2} - \frac{z}{2})\Gamma(\frac{1}{2} + \frac{z}{2})}{\Gamma(1 + \frac{v}{2} - \frac{z}{2})} \left(\frac{x}{2}\right)^{-z} dz,$$

$$\max\{-1, -\mathrm{Re}\{v\}\} < \delta < 1. \tag{7.3}$$

The change of variable $z/2 = w$ yields the following expression as a G-function:

$$f_v(x) = \frac{1}{2}\, G^{21}_{13}\left(\frac{x^2}{4}\,\middle|\, \begin{matrix} \frac{1}{2} \\ \frac{1}{2} & \frac{v}{2} & -\frac{v}{2} \end{matrix}\right). \tag{7.4}$$

We will not dwell on this G-function answer, which, once again, is not very revealing, and for which no simplification is provided in the "reduction table" [6, 8.4.52]. For general v, and to the left of the integration path in (7.3), there are two semi-infinite lattices of *simple* poles. (Note that double poles arise in the limit $v \to 2m + 1$; we ignore this for now.) Closing the contour at left (this is justified in Chapter 8) and calculating the residues, one obtains an expression involving two power series

$$f_v(x) = \frac{x}{4}\sum_{n=0}^{\infty}\frac{\Gamma(\frac{v}{2} - \frac{1}{2} - n)}{\Gamma(\frac{v}{2} + \frac{3}{2} + n)}\left(-\frac{x^2}{4}\right)^n$$

$$+ \frac{1}{2}\left(\frac{x}{2}\right)^v\sum_{n=0}^{\infty}\frac{\Gamma(\frac{1}{2} + \frac{v}{2} + n)\Gamma(\frac{1}{2} - \frac{v}{2} - n)}{n!\,\Gamma(1 + v + n)}\left(-\frac{x^2}{4}\right)^n. \tag{7.5}$$

The gamma functions are of the form $\Gamma(z + n)$ or $\Gamma(z - n)$ and can be expressed in terms of Pochhammer's symbols using (2.26) or (2.31), respectively. The resulting series can be immediately identified with $_pF_q$'s, so that

$$f_v(x) = \frac{x}{4}\frac{\Gamma(\frac{v}{2} - \frac{1}{2})}{\Gamma(\frac{v}{2} + \frac{3}{2})}\,_1F_2\left(1; \frac{3}{2} + \frac{v}{2}, \frac{3}{2} - \frac{v}{2}; \frac{x^2}{4}\right)$$

$$+ \frac{1}{2}\left(\frac{x}{2}\right)^v\frac{\Gamma(\frac{1}{2} + \frac{v}{2})\Gamma(\frac{1}{2} - \frac{v}{2})}{\Gamma(1 + v)}\,_0F_1\left(1 + v; \frac{x^2}{4}\right). \tag{7.6}$$

This time, further simplification is possible using the aforementioned "reduction tables" of $_pF_q$. Specifically, from [6, Entry 7.14.3.6] and [6, Entry 7.13.1.1] one obtains

$$f_v(x) = is_{0,v}(ix) + \frac{\pi}{2}\frac{1}{\cos(\pi v/2)}I_v(x), \tag{7.7}$$

in which I_v is the usual modified Bessel function and $s_{0,v}$ is the Lommel function discussed in [19, 8.57] (see also [50]) or [6, II.12]. Like all manipulations in our previous examples, those used up to now (to obtain (7.7) from (7.2)) have been straightforward.

The answer (7.7) is simple enough, but both terms become infinite in the case $v = 1, 3, \ldots,$* which is precisely the case we are interested in. Since the original integral (7.1) is

---

* This is obvious for the second term in (7.7); for the first term, see the definition of $s_{0,v}$ in [19, Entry 8.570.1] or [6, II.12].

finite, the two terms in (7.7) must combine to give a quantity that remains finite in the limit $\nu \to 2m + 1$. To calculate this quantity, use [19, Entry 8.570.2] to express $s_{0,\nu}$ in terms of J_ν, Y_ν and the Lommel function $S_{0,\nu}$. Then, combine the J_ν and Y_ν with the I_ν of (7.7) using the Bessel-function identities [51, Entries 9.1.2, 9.6.2, and 9.6.3]

$$Y_\nu(z) = \frac{1}{\sin \pi \nu} [\cos \pi \nu J_\nu(z) - J_{-\nu}(z)], \tag{7.8}$$

$$J_\nu(z e^{i\pi/2}) = e^{i\pi\nu/2} I_\nu(z), \qquad -\pi < \arg z \le \pi/2, \tag{7.9}$$

and

$$K_\nu(z) = -\frac{\pi}{2 \sin \pi \nu} [I_\nu(z) - I_{-\nu}(z)], \tag{7.10}$$

as well as properties of the gamma function. One can readily arrive at

$$f_\nu(x) = i S_{0,\nu}(ix) - i e^{-i\nu\pi/2} K_\nu(x), \tag{7.11}$$

which is an alternative to (7.7) expression for general ν. Clearly, the right-hand side of (7.11) is finite when $\nu \to 2m + 1$. Furthermore, by [19, Entry 8.573.2] or [6, Entry II.12], in the limit, $S_{0,\nu} = S_{0,2m+1}$ reduces to a polynomial, a formula for which is provided in [19, Entry 8.590.1] or [6, Entry II.24]. Using that formula yields our final expression for $f_{2m+1}(x)$:

$$f_{2m+1}(x) = \frac{1}{2} \sum_{n=0}^{m} (-1)^n \frac{(m+n)!}{(m-n)!} \left(\frac{2}{x}\right)^{2n+1} + (-1)^{m+1} K_{2m+1}(x). \tag{7.12}$$

This simple answer consists of a polynomial of degree $2m + 1$ in $1/x$, containing odd powers of $1/x$ only, plus/minus a modified Bessel function. Equation (7.12) is excellent both for numerical evaluation and for further analytical work. This is especially true for *large* values of x, where $K_{2m+1}(x)$ is very small and the polynomial strongly dominates. (Large x is of interest in [44], [45].)

It is very instructive to return to (7.3), set $\nu = 2m + 1$ ($m = 0, 1, \ldots$), and discuss the Mellin–Barnes integral

$$f_{2m+1}(x) = \frac{1}{4} \frac{1}{2\pi i} \int_{\delta-i\infty}^{\delta+i\infty} \frac{\Gamma(m + \frac{1}{2} + \frac{z}{2})\Gamma(\frac{1}{2} - \frac{z}{2})\Gamma(\frac{1}{2} + \frac{z}{2})}{\Gamma(m + \frac{3}{2} - \frac{z}{2})} \left(\frac{x}{2}\right)^{-z} dz, \qquad -1 < \delta < 1, \tag{7.13}$$

directly for the special case, in which some of the simple poles coalesce, yielding double poles: In the special case, to the left of the contour one has a *finite number of simple poles* (at $z = -1, -3, \ldots, -(2m-1)$) and, after that, a *semi-infinite lattice of double poles* (at $z = -(2m+1), -(2m+3), \ldots$). For $m = 2$, this is illustrated in Table 7.1. Despite its intricate appearance, this pole structure has much to reveal. Consider, for example, closing the contour

TABLE 7.1: Poles of the Integrand of (7.13) when $m = 2$

−11	−9	−7	−5	−3	−1	1	3	5	7	9	11
...	PP	PP	PP	P	P	P	P	P			

The thick line between −1 and 1 indicates that the inversion path lies in the strip $-1 < \text{Re}\{z\} < 1$. The two simple poles (P) to the left of the contour are due to $\Gamma(\frac{1}{2} + \frac{z}{2})$. The semi-infinite lattice of double poles (PP) which starts at $z = -5$ is due to $\Gamma(m + \frac{1}{2} + \frac{z}{2})\Gamma(\frac{1}{2} + \frac{z}{2})$, which in this case equals $\Gamma(\frac{5}{2} + \frac{z}{2})\Gamma(\frac{1}{2} + \frac{z}{2})$. Note that there are three poles to the right of the contour; these stop at $z = 5$ because the contributions (poles) from $\Gamma(\frac{1}{2} - \frac{z}{2})$ in the numerator are canceled by the contributions (zeros) from $\Gamma(m + \frac{3}{2} - \frac{z}{2})$ (equal, in this case, to $\Gamma(\frac{7}{2} - \frac{z}{2})$) in the denominator. For general integer m, there are m simple poles to the left of the contour, followed by a semi-infinite lattice of double poles and, also, $m + 1$ poles to the right of the contour.

at left and evaluating residues. The expression obtainable in this manner consists of (i) two power series in x, one of which is multiplied by $\ln x$ (arising from the double poles, just as in Chapter 6), plus (ii) a polynomial in x (arising from the simple poles). This expression is complicated in form; in any case, it is an ascending series, in which no negative powers of x appear.

Therefore, if one expands the right-hand side of the answer (7.12) in an ascending series, the negative powers must cancel out! This prediction, which can be easily verified (use the well-known ascending series [51, Entry 9.6.11] for $K_{2m+1}(x)$), signifies that much information can be obtained simply by glancing at the pole structure of Mellin–Barnes integrals.

As illustrated for $m = 2$ in Table 7.1, in addition to the poles to the left of the contour, there is also is a finite number of *simple* poles to the *right* of the contour. We will briefly comment on their significance in Section 9.4.

CHAPTER 8

On Closing the Contour

Within the context of G-functions, certain sufficient conditions enabling one to "close a vertical contour at left" were discussed in Section 3.1. In this section, we present further such conditions. More precisely, one starts from a *convergent* Mellin–Barnes integral and transforms the original, vertical contour to one surrounding all poles to the left of the contour, and closing at $-\infty$. This is possible if the integrals over two large quarter-circles vanish, something that can be systematically investigated via Stirling's formula. The following very simple set of *sufficient* conditions that arise in this manner can be found in [13, Chapt. 5] or [14] and is adequate for the purposes of this book.

We discuss the Mellin–Barnes integral

$$\frac{1}{2\pi i} \int_{\delta-i\infty}^{\delta+i\infty} \frac{\prod_{l=1}^{m} \Gamma(a_l + A_l z) \prod_{l=1}^{n} \Gamma(b_l - B_l z)}{\prod_{l=1}^{p} \Gamma(c_l + C_l z) \prod_{l=1}^{q} \Gamma(d_l - D_l z)} \, x^{-z} \, dz, \tag{8.1}$$

where A_l, B_l, C_l, and D_l are strictly positive. Define the quantities

$$\Delta = \sum_{l=1}^{m} A_l + \sum_{l=1}^{q} D_l - \sum_{l=1}^{n} B_l - \sum_{l=1}^{p} C_l \tag{8.2}$$

and

$$\Pi = -\ln|x| + \sum_{l=1}^{m} A_l \ln A_l + \sum_{l=1}^{q} D_l \ln D_l - \sum_{l=1}^{n} B_l \ln B_l - \sum_{l=1}^{p} C_l \ln C_l. \tag{8.3}$$

As x varies, Π will take on all real values. One has the following cases:

Case 1: If $\Delta > 0$, one can close the contour at *left* for all real x.

Case 2: If $\Delta < 0$, one can close the contour at *right* for all real x.

Case 3: If $\Delta = 0$ one can close at left if $\Pi > 0$ and at right if $\Pi < 0$.

A discussion of the more complicated case $\Delta = \Pi = 0$ can be found in [13] and [14].

The examples in Section 4.2 and Chapters 5–7 fall under Case 1, with $\Delta > 0$, so we closed at left and were subsequently lead to convergent series in ascending powers of x. Cases 2 and 3 also lead to convergent series representations. Case 3 arises in Section D.2 of Appendix D, which concerns an integral arising when studying the solvability of certain integral equations for the current on a thin-wire loop antenna [52].

For more complete discussions of closing the contour and subsequently obtaining $f(x)$ from $\tilde{f}(z)$, the interested reader can look up "Slater's theorem" in [7].

CHAPTER 9

Further Discussions

9.1 A NOTE REGARDING MELLIN CONVOLUTION

Within the context of the Mellin transform, we defined convolution by (2.12) and called it "Mellin convolution." This is not standard; another definition often encountered is $\int_0^\infty h(y)g(x/y)/y\,dy$ (see, for example, [3], [6], [7], [9], [12], [42]). The advantage of this somewhat more complicated definition is that the Mellin transform of the convolution is the product of the two individual Mellin transforms; as opposed to (2.13), no translation or reflection is required.

9.2 ON THE USE OF SYMBOLIC ROUTINES

It has already been mentioned that the Mellin-transform method is important for Mathematica's symbolic routine `Integrate[]`. Symbolic routines can also help when one applies the Mellin-transform method on his/her own, as they can be used in many intermediate steps. Such steps include the "lookup" of Mellin transforms, "messy" manipulations such as the calculation of residues, and the simplification of complicated expressions. Symbolic routines are powerful tools and it pays to be flexible when using them. When applied to the right-hand side of (7.4), for example, Mathematica 5.0's routine `FullSimplify[]` *does not* yield (7.6), even when $x > 0$ is assumed. When applied to the right-hand side of (7.6) minus the right-hand side of (7.4), however, `FullSimplify[]` *does* yield zero. Here, `FullSimplify[]` verifies the answer but cannot produce it from scratch.

9.3 COMPLEX VALUES OF THE PARAMETER x

When applying the Mellin-transform method, we always took $x > 0$. Often, integrals to be evaluated can be written as in (4.1) but with complex x. A standard way to proceed is to assume initially that $x > 0$, proceed with the Mellin-transform method, and attempt to analytically continue the final answer to complex x. We apply a related idea when evaluating the integral of Section D.2 of Appendix D: there, temporarily changing the interval of interest (from $x > 2$ to $0 < x < 1$) allows us to close the contour at left.

9.4   SIGNIFICANCE OF THE POLES TO THE RIGHT; ASYMPTOTIC EXPANSIONS

What is the significance of poles to the *right* of the contour? As the reader might have guessed, picking up residues from these poles usually leads to *descending* series, which are typically large-x expansions (asymptotic expansions for $x \gg 1$) and which often diverge. (Note, however, that one has convergence in Cases 2 and 3 of Chapter 8.)

Asymptotic expansions are important for applications, but are beyond the scope of this book. It is nonetheless worth commenting on the significance of the $m + 1$ poles to the right of the contour in the example of Chapter 7. See Table 7.1 for an illustration of the $m = 2$ case. It can be shown that the $m + 1$ inverse powers of x arising from these poles indeed form an asymptotic expansion of $f_{2m+1}(x)$ for large x. This asymptotic expansion has the peculiar property of having a *finite* number of terms.

Recall that any function has at most one asymptotic expansion involving inverse powers of x and that such an expansion remains unaltered if an exponentially small quantity is added to the original function. Since $K_{2m+1}(x)$ is exponentially small for large x, the aforementioned $m + 1$ inverse powers of x must coincide with the inverse powers of x in the answer (7.12)! This is indeed the case, so we have a second prediction (different from the one in Chapter 7) illustrating that Mellin–Barnes integrals can be very informative.

9.5   RELATIONS OF OUR RESULTS TO ENTRIES IN INTEGRAL TABLES

Here, we discuss which of our results can be found in the standard integral tables [19] and [4]–[6].

(i) *Integral of Section 4.2:* It appears that neither [4] nor [19] contain an evaluated form for the integral (4.4) of Section 4.2. Note, in particular, that if one writes $\sin^2 xy = (1 - \cos 2xy)/2$, one cannot split the integral because the resulting two integrals diverge (see Rule 1 of Appendix A). Thus, the tabulated integrals [4, Entry 2.5.8.3] and [19, Entry 3.771.4] are not helpful for the integral (4.4).

(ii) *Integral in Chapter 5:* The evaluated form in (5.7) is a simple consequence of the tabulated integral [5, Entry 2.12.32.3], which is more general.

(iii) *Integral in Chapter 6:* Neither [4] nor [19] appear to contain an evaluated form for the integral (6.2) of Chapter 6. Note that if one writes $\sin^2 xy = (1 - \cos 2xy)/2$, the tabulated integrals [4, Entry 2.5.8.3] and [4, Entry 2.5.2.2] are not *directly* applicable to our $f(x)$ because the forms there are, in our case, indeterminate.

(iv) *Integral in Chapter 7:* With regard to the example in Chapter 7, to the best of the author's knowledge, (7.12) is not available in the literature (apart from certain special cases), whereas both (7.7) and (7.11) can be found in the corrected by the author [50] tabulated integral [19, Entry 6.532.1]. The form (7.7) also appears as [5, Entry 2.12.4.20].

9.6   NUMERICAL EVALUATION OF INTEGRALS BY MODERN ROUTINES

One may have the view that, like the modern routines for the $_pF_q$, modern numerical-integration routines can themselves often be used as black boxes. We discuss this by focusing on integrals like those in Chapters 6 and 7, which have an infinite upper integration limit, and oscillations due to a factor $\cos xy$ or $J_\nu(xy)$ in the integrand. Let us consider three relevant routines (or types of routines) whose use seems to be widespread:

(i) As far as accuracy is concerned, the best routine the author is aware of is Mathematica's NIntegrate[], provided the Method->Oscillatory option is used. The user sets the upper integration limit to Infinity. To give an example, one gets highly accurate results by evaluating (7.1) in this manner; but the computer time is significantly less if one uses (7.12).

(ii) Even for the types of integrals discussed here, it is not always possible to use the Method->Oscillatory option, e.g., when the ν in $J_\nu(xy)$ is negative or complex. In such cases, a Mathematica user can resort to NIntegrate[] without the Method->Oscillatory option; the integration limit can still be set to Infinity. The accuracy in such cases is significantly less than before. For example, when $x = 3$ and $\nu = -0.9$ in (7.2), one gets a result correct to within only 0.3%, accompanied by a warning message. By contrast, (7.6) quickly yields answers which, as far as the author can tell, are highly accurate.

(iii) The numerical integration routines in Matlab 7.0 (quad, quadl) do not allow the user to specify an infinite integration limit. With such types of routines, one often specifies an integration limit large enough to yield a desired accuracy. For slowly decaying integrands, such "truncation methods" may not work well at all. As an example, consider the integral $\int_1^\infty y^{-\nu} \cos xy \, dy$ ($\nu > 0$, $x > 0$) which, as the reader may wish to verify, can be evaluated in terms of a $_1F_2$. A truncation method amounts to numerically integrating $\int_1^M y^{-\nu} \cos xy \, dy$. For $\nu = 1/2$, both the integral and its truncated version can be evaluated in terms of Fresnel integrals [51, Chapt. 7], so, for this case, we can compute the best one can do by *any* truncation method. As it turns out, the required values of M are very, very large: For $x = 2$, a value $M = 12\,000$ is necessary for 1% accuracy, while a value $M = 48\,000$ is necessary for 0.5% accuracy. For either value of

M, an actual numerical-integration routine will certainly provide much less accuracy. The situation deteriorates even more if ν decreases.

9.7 ADDITIONAL READING

In this section, we give some further references to the topics of this book.

References [12], [53], and [54] contain introductory treatments of the Mellin transform. They include brief descriptions of the Mellin-transform method (none of these discuss G, while $_pF_q$ is only discussed, very briefly, in [54]), as well as short discussions of (and references to) other applications of the Mellin transform.

Gamma and related functions are treated in most textbooks on complex variables and special functions. Besides [15] (which we referred to several times in Chapter 2), we mention [11], [16], and [55]. Often, such textbooks also discuss, in detail, the so-called Gauss hypergeometric function $_2F_1$, to which the $_pF_q$ is a generalization.

Reference [16] discusses the gamma function, the $_pF_q$, and the G-function, and includes derivations. More extensive references for these topics are [21] and [56]. Besides [6], many relevant formulas can be found in [19] and online at [20]. Reference [57] gives one an idea of how G-functions are handled by modern symbolic programs.

On Mellin–Barnes integrals, see [7], [16], [56], and the comprehensive book [58]. Reference [59] is a monograph on the Fox H-function and its extensions to more than one variable.

References [9] and [12]–[14] contain simple, informative discussions relevant to the Mellin-transform method, not too different from the general material in Chapters 2–4. More detailed expositions can be found in the pioneering (but readable) works [3], [7]. We note that [3] cites over 1600 references! As discussed previously, the most comprehensive lists of formulas required for the application of the Mellin-transform method are contained in the reference work [6], which has no derivations.

Many books on asymptotic expansions have little to say on Mellin transforms; for exceptions, see the brief discussions in [60] and [61], as well as the comprehensive works [58] and [62].

The origins of what we call the "Mellin-transform method" go far back: The idea of the Mellin inversion formula appeared in an 1876 memoir by Riemann, and the first accurate discussion was given by Mellin in 1896 and 1902. What we now call "Mellin–Barnes integrals" were first introduced by Pincherle in 1888, developed theoretically by Mellin by 1910, and used by Barnes in 1908 to discuss the asymptotic expansion of certain special functions. The definition and a first systematic study of the G-function appeared in 1936 in a paper by Meijer. The pioneering work of Pincherle is described in [63], while biographies of Mellin and Barnes appear in [58, Chapt. 1].

C H A P T E R 10

Summary and Conclusions

What we have called the "Mellin-transform method" is an extremely powerful technique for the exact evaluation of definite integrals. It can often be combined with other methods and is applicable to a wide class of integrals. It is a significant constituent of certain modern symbolic integration packages, and has been employed in an essential manner to compile what may be the most comprehensive published table of integrals [4]–[6]. It is not as widely known as it should be. In many cases, the method is completely straightforward, yielding results difficult to find elsewhere or to derive by alternative methods.

When applicable, the Mellin-transform method typically yields ascending series (which often involve logarithms or powers of logarithms) or expressions involving the generalized hypergeometric function $_pF_q$ or Meijer's G-function. Because such expressions can be automatically handled by modern numerical routines, they are much more useful than in the past. Because the $_pF_q$ and G possess a vast number of documented properties, such expressions can also be a good first step for further analytical work. Often, though, expressions involving $_pF_q$ and G are merely an intermediate step as they can be simplified by lookup in extensive tables or by symbolic routines.

To apply the method, one should have some familiarity with the $_pF_q$, Mellin–Barnes integrals, and the G-function and possess some experience with certain lookup tables. More importantly, one should have a good working knowledge of the basics of the Mellin transform, as well as of gamma and related functions. All these topics, which can be understood by one familiar with complex analysis, are discussed in Chapters 2 and 3 of this book. These chapters place little emphasis on mathematical details or fine points. Section 9.7 is a guide to additional literature on the aforementioned topics and to literature on the Mellin-transform method itself. A lemma useful for "messy" calculations, sometimes required when applying the Mellin-transform method, is included in Section 2.7.

We illustrated the method by treating four example-integrals (Section 4.2 and Chapters 5–7), all arising from antenna/electromagnetics problems. Some of our answers do not appear even in the standard integral tables. All answers are suitable for numerical evaluation and most are believed to be new, at least in the antenna/electromagnetics literature. Two of the answers lead, additionally, to simple approximate formulas for the integral that significantly improve upon formulas of standard antenna textbooks. We have thus explicitly illustrated highly desirable features of the Mellin-transform method in the specific context of electromagnetics and antenna theory.

Appendix A: On the Convergence/Divergence of Definite Integrals

In this stand-alone appendix, we discuss simple methods for investigating whether a given definite integral converges or diverges. We specifically consider what mathematicians call "improper integrals" (we omit the adjective "improper" throughout this book) in which the integration interval extends to infinity or the integrand is infinite at finite points within the integration interval.

It is possible to have a convergent integral when the integrand is infinite at finite points; but the infinity cannot be too rapid. For an infinite integration interval, it is *not* sufficient for the integrand to vanish at infinity for the integral to converge*; the rate of this vanishing is also important. In this appendix, we provide simple, systematic rules of thumb that can help us determine when a given integral converges and apply these rules to a large number of examples. Our integration path is always on the real axis.

A.1 SOME REMARKS ON OUR RULES

Before giving our rules, some clarifications are necessary.

(i) In this appendix, and throughout this book, our integrals are "classical" ones. Thus we consider integrals such as $\int_{-\infty}^{\infty} \exp(ixy)\, dy$ to be divergent because the double limit

$$\frac{1}{ix}\left(\lim_{M \to \infty} e^{ixM} - \lim_{N \to -\infty} e^{ixN}\right) \qquad (A.1)$$

does not exist. A meaning to such integrals can of course be attached with the aid of the Dirac delta function (the above integral is then equal to $2\pi\,\delta(x)$). Alternatively, one can attach meaning via "Abel summability"; see [64]. In this book, however, we regard the integral as divergent because we choose to use a different mathematical framework.

---

* This is obvious, for example, from Rule 3 later. Conversely, if the integral converges, it is not necessary that the integrand vanishes at infinity. This is seen from the Fresnel integral $\int_0^{\infty} \cos(\pi t^2/2)\, dt$, the convergence of which can be shown by changing the variable $x = \pi t^2/2$ and using Rule 5 later.

(ii) In what follows, we are interested in the behavior of the integrand only at one endpoint of the integration interval, 0 or $+\infty$. Thus, we examine integrals of the type $\int_0^A$ and $\int_A^\infty$ where $A > 0$. In the remaining integration interval, we assume that our integrands are "sufficiently smooth," a property we make no attempt to clarify further.

(iii) We also assume that the convergence/divergence of the integral can be determined if we replace the integrand by its leading asymptotic approximation. For a "sufficiently small" integration interval—that is, for sufficiently small (large) A for integrals of the form $\int_0^A$ ($\int_A^\infty$)—such a replacement is usually legitimate.

(iv) Because we do not precisely specify the above-mentioned limitations on the integrand and the integration interval, the analysis that follows is not rigorous. Despite this, the convergence/divergence rules that follow are very useful in practice. More rigorous treatments can be found in the literature [6, Appendix I], [64]. A historical account of relevant topics can be found in [65].

A.2 RULES FOR DETERMINING CONVERGENCE/DIVERGENCE

We now present our rules, together with justifications.

Rule 1—Algebraic behavior at zero: If

$$g(y) \sim \frac{B}{y^{1-\epsilon}} \quad \text{as} \quad y \to 0^+ \ (B \neq 0), \tag{A.2}$$

then the integral $\int_0^A g(y)\,dy$

(i) converges if $\mathrm{Re}\{\epsilon\} > 0$ and

(ii) diverges if $\mathrm{Re}\{\epsilon\} \leq 0$.

Justification of Rule 1: The rule is valid because

$$\int_0^A \frac{dy}{y^{1-\epsilon}} = \frac{A^\epsilon}{\epsilon} - \lim_{y \to 0+} \frac{y^\epsilon}{\epsilon} \tag{A.3}$$

(here, we assumed $\epsilon \neq 0$) and the limit in (A.3) exists and is finite only when $\mathrm{Re}\{\epsilon\} > 0$. For the special case $\epsilon = 0$, we have

$$\int_0^A \frac{dy}{y} = \ln A - \lim_{y \to 0+} \ln y, \tag{A.4}$$

in which the limit does not exist, so that the integral diverges.

Rule 2 that follows is a generalization of Rule 1, allowing a logarithmic dependence to multiply the algebraic one.

Rule 2—Algebraic/logarithmic behavior at zero: If δ is real and

$$g(y) \sim B\frac{(-\ln y)^{\delta}}{y^{1-\epsilon}} \quad \text{as} \quad y \to 0^{+} \ (B \neq 0), \tag{A.5}$$

then the integral $\int_{0}^{A} g(y)\, dy$

    (i) converges if $\text{Re}\{\epsilon\} > 0$ and

    (ii) diverges if $\text{Re}\{\epsilon\} < 0$.

In other words, at least for $\text{Re}\{\epsilon\} \neq 0$, the presence of the logarithm does not affect the convergence or divergence of the integral.

Justification of Rule 2: With a change of variable $-\ln y = t$, we see that

$$\int_{0}^{A} \frac{(-\ln y)^{\delta}}{y^{1-\epsilon}}\, dy = \int_{-\ln A}^{\infty} e^{-\epsilon t} t^{\delta}\, dt, \tag{A.6}$$

and it is obvious that the integral in (A.6) converges if $\text{Re}\{\epsilon\} > 0$ and diverges if $\text{Re}\{\epsilon\} < 0$. (Using the above change of variable, it is also possible to come up with a rule for the case $\text{Re}\{\epsilon\} = 0$, but the said rule is more complicated and depends on δ.)

Rule 3—Algebraic behavior at infinity: If

$$g(y) \sim \frac{B}{y^{1+\epsilon}} \quad \text{as} \quad y \to +\infty \ (B \neq 0), \tag{A.7}$$

then the integral $\int_{A}^{\infty} g(y)\, dy$

    (i) converges if $\text{Re}\{\epsilon\} > 0$ and

    (ii) diverges if $\text{Re}\{\epsilon\} \leq 0$.

Justification of Rule 3: Similar to that of Rule 1.

    The following rule generalizes Rule 3.

Rule 4—Algebraic/logarithmic behavior at infinity: If δ is real and

$$g(y) \sim B\frac{(\ln y)^{\delta}}{y^{1+\epsilon}} \quad \text{as} \quad y \to \infty \ (B \neq 0), \tag{A.8}$$

then the integral $\int_0^A g(y)\,dy$

    (i) converges if $\mathrm{Re}\{\epsilon\} > 0$ and

    (ii) diverges if $\mathrm{Re}\{\epsilon\} < 0$.

Here, as in Rule 2, for $\mathrm{Re}\{\epsilon\} \neq 0$ the presence of the logarithm does not affect the convergence or divergence.

Justification of Rule 4: Similar to that of Rule 2, with a change of variable $\ln y = t$.

Rule 5—Sinusoidal/algebraic behavior at infinity: If x is real with $x \neq 0$ and

$$g(y) = O\left(\frac{1}{y^\epsilon}\right) \quad \text{as} \quad y \to \infty, \tag{A.9}$$

then the integrals

$$\int_A^\infty g(y) \cos xy\,dy \quad \text{and} \quad \int_A^\infty g(y) \sin xy\,dy \tag{A.10}$$

    (i) converge if $\mathrm{Re}\{\epsilon\} > 0$ and, in particular,

    (ii) they converge absolutely if $\mathrm{Re}\{\epsilon\} > 1$.

(Recall that the integral $\int_A^\infty g(y)\,dy$ converges absolutely if $\int_A^\infty |g(y)|\,dy$ converges.) Comparing with Rule 3, we see that the condition on $g(y)$ for convergence is weaker; the convergence for $0 < \mathrm{Re}\{\epsilon\} < 1$ is due to the *oscillations* of the integrand.

Justification of Rule 5: From the inequalities

$$\left|\frac{\sin xy}{y^\epsilon}\right| \leq \frac{1}{y^{\mathrm{Re}\{\epsilon\}}}, \qquad \left|\frac{\cos xy}{y^\epsilon}\right| \leq \frac{1}{y^{\mathrm{Re}\{\epsilon\}}}, \tag{A.11}$$

and Rule 3, the absolute convergence (and therefore, the convergence) for $\mathrm{Re}\{\epsilon\} > 1$ follows immediately. For $\mathrm{Re}\{\epsilon\} > 0$, an integration by parts yields

$$\int_A^\infty \frac{\sin xy}{y^\epsilon}\,dy = -\frac{1}{x}\lim_{y\to\infty}\frac{\cos xy}{y^\epsilon} + \frac{1}{x}\frac{\cos Ax}{A^\epsilon} - \frac{\epsilon}{x}\int_A^\infty \frac{\cos xy}{y^{1+\epsilon}}\,dy. \tag{A.12}$$

In the right-hand side of (A.12), the limit exists and is zero, while the last integral converges (absolutely) as we just showed. Thus, for $\mathrm{Re}\{\epsilon\} > 0$, the integral $\int_A^\infty \frac{\sin xy}{y^\epsilon}\,dy$ converges. The convergence, for $\mathrm{Re}\{\epsilon\} > 0$, of the integral $\int_A^\infty \frac{\cos xy}{y^\epsilon}\,dy$ can be established in a similar manner.

A.3   EXAMPLES

Applying the above rules, the reader can verify that the following integrals converge:

$$\int_1^\infty \frac{y^2}{\sqrt{y^7 + 1}} \, dy, \tag{A.13}$$

(the integral $\int_1^\infty \frac{y^3}{\sqrt{y^7+1}} \, dy$, however, diverges),

$$\int_0^\infty \frac{\sin y}{\sqrt{y}} \, dy, \tag{A.14}$$

$$\int_0^\infty \frac{\sin y}{y} \, dy, \tag{A.15}$$

(but $\int_0^\infty \frac{\cos y}{y} \, dy$ diverges),

$$\int_1^\infty \sin\left(\frac{1}{y^2}\right) dy, \tag{A.16}$$

(the integral in (A.16) converges absolutely),

$$\int_0^\infty \frac{\cos xy}{\sqrt{1 + y^2}} \, dy, \qquad x > 0, \tag{A.17}$$

$$\int_0^\infty K_0(y) \cos xy \, dy, \qquad x > 0, \tag{A.18}$$

$$\int_0^\infty y^{-1/2} K_0(y) \cos xy \, dy, \qquad x > 0, \tag{A.19}$$

$$\int_0^\infty \frac{y^{z-1}}{\sqrt{1 + y^2}} \, dy, \qquad 0 < \text{Re}\{z\} < 1, \tag{A.20}$$

$$\int_0^\infty \frac{y^{z-1}}{e^y - 1} \, dy, \qquad \text{Re}\{z\} > 1, \tag{A.21}$$

$$\int_0^\infty \frac{\sin y}{1 + y^2} y^{z-1} \, dy, \qquad -1 < \text{Re}\{z\} < 3, \tag{A.22}$$

$$\int_0^\infty y^{z-1} \ln\left(y + \sqrt{y^2 + 1}\right) dy, \qquad -1 < \text{Re}\{z\} < 0. \tag{A.23}$$

The last four integrals are of course Mellin transforms, with each restriction on z denoting the corresponding strip of analyticity (SOA), as defined in Section 2.1. The reader can also use our rules to verify the SOA's of Table 2.1, as well as the convergence of the integrals (4.4), (5.4) (subject to the restrictions in (5.5)), (6.2), (7.1), (7.2), and (D.1) below.

Appendix B: The Lemma
of Section 2.7

In this appendix we show (2.38), which is the lemma (Application 5) of Section 2.7. Our derivation is based on a number of identities—interesting in their own right—involving $\Gamma(z)$, $\psi(z)$, and $(z)_n$.

B.1 PRELIMINARY IDENTITIES

Our first identity, which is a useful expression for the derivative of $(z)_n$,

$$\frac{d}{dz}(z)_n = (z)_n \left[\psi(z+n) - \psi(z)\right], \qquad (B.1)$$

is a simple consequence of the definitions (2.22) and (2.26) of $\psi(z)$ and $(z)_n$. The identity

$$\psi(z+n) - \psi(z) = \sum_{l=0}^{n-1} \frac{1}{z+l}, \qquad n = 0, 1, 2, \dots, \qquad (B.2)$$

can be verified by induction on n and use of the recurrence formula (2.23) for the psi function. Taking the limit $z \to -n$ in (B.2) yields

$$\lim_{z \to -n} \left[\psi(z+n) - \psi(z)\right] = -\gamma - \psi(n+1), \qquad n = 0, 1, 2, \dots, \qquad (B.3)$$

where (2.24) and (2.25) were used. The identity

$$(-n)_n = (-1)^n n!, \qquad n = 0, 1, 2, \dots, \qquad (B.4)$$

can be shown by setting $z = n+1$ in (2.31) and using (2.17). Finally, from the definition (2.26) of $(z)_n$ and the recurrence formula (2.16), it is apparent that

$$\Gamma(z) = \frac{1}{z+n} \frac{\Gamma(z+n+1)}{(z)_n}, \qquad n = 0, 1, 2, \dots. \qquad (B.5)$$

By (B.4) and $\Gamma(1) = 1$, the quantity multiplying $\frac{1}{z+n}$ in (B.5) is finite and nonzero when $z = -n$. Thus, (B.5) shows explicitly that $\Gamma(z)$ has a simple pole at $z = -n$ (in Section 2.4, we showed this via (2.15)) and is convenient for residue calculations.

B.2   DERIVATION OF (2.38)

To determine the residue at $z = 0, -1, -2, \ldots$ of the quantity $h(z)$ defined by

$$h(z) = [\Gamma(z)]^2 g(z) x^{-z}, \qquad (B.6)$$

use (B.5) to write

$$h(z) = \frac{1}{(z+n)^2} \left[\frac{\Gamma(z+n+1)}{(z)_n} \right]^2 g(z) x^{-z}, \qquad n = 0, 1, 2, \ldots . \qquad (B.7)$$

Since $g(z)$ is analytic and nonzero at $z = -n$, (B.7) explicitly shows that $h(z)$ has a double pole at $z = -n$ and allows one to find the residue in the usual manner by differentiating $(z + n)^2 h(z)$ with respect to z and then setting $z = -n$. This leads to

$$\mathrm{Res}\,\{h(z); z = -n\} = \left[\frac{\Gamma(z+n+1)}{(z)_n} \right]^2 x^{-z}$$

$$\times \ \{-g(z) \ln x + 2\psi(z + n + 1)g(z) - 2[\psi(z + n) - \psi(z)]g(z) + g'(z)\} \,\big|_{z=-n}, \qquad (B.8)$$

where (B.1) was used. The desired eqn. (2.38) then follows from (B.8), (B.3), (B.4), and (2.25).

Appendix C: Alternative Derivations or Verifications for the Integrals of Section 4.2, and Chapters 5 and 6

In this appendix, we provide alternative derivations, that do not make use of the Mellin-transform method, for the integrals in Section 4.2 and Chapters 5 and 6. For the integral in Chapter 6, the procedure is more of an after-the-fact verification than a derivation from scratch.

(i) *Integral of Section 4.2:* To show (4.10) from (4.4), expand $\sin^2 xy/(xy)^2$ into its Taylor series (this can be done by writing $\sin^2 xy = (1 - \cos 2xy)/2$ and using the well-known Taylor series for the cosine). Then, integrate term by term using Entry 2 of Table 2.1, which is a standard tabulated integral.

(ii) *Integral of Chapter 5:* Similarly, to come up with the series form corresponding to (5.7), start from (5.4) and expand $J_\mu(xy)J_\nu(xy)$ into its Taylor series, which is provided in [19, Entry 8.442.1]. This series is

$$J_\mu(xy)J_\nu(xy) = \sum_{n=0}^{\infty} \frac{(-1)^n}{n!} \frac{\Gamma(\nu + \mu + 2n + 1)}{\Gamma(\nu + \mu + n + 1)\Gamma(\nu + n + 1)\Gamma(\mu + n + 1)}$$
$$\times \left(\frac{xy}{2}\right)^{\nu+\mu+2n}. \qquad (C.1)$$

Then, integrate term by term using Entry 2 of Table 2.1. The derivation we just outlined can be found in [30].

(iii) *Integral of Chapter 6:* To verify (6.5), start from (6.2), write $\sin^2 xy = (1 - \cos 2xy)/2$, and split the integral. Evaluate the first integral, which is elementary, to arrive at

$$f(x) = \frac{1}{2x^2}[1 - g(x)], \qquad (C.2)$$

where

$$g(x) = \int_1^\infty \frac{\cos 2xy}{y^2\sqrt{y^2 - 1}} \, dy. \qquad (C.3)$$

To evaluate the integral $g(x)$, twice differentiate (C.3) with respect to x. The resulting integral is readily recognized from

$$Y_0(2x) = -\frac{2}{\pi} \int_1^\infty \frac{\cos 2xy}{\sqrt{y^2 - 1}} \, dy, \qquad x > 0, \qquad (C.4)$$

which is a standard integral representation of Y_0 [51, Entry 9.1.24]. It follows that $g(x)$ satisfies the differential equation

$$g''(x) = 2\pi Y_0(2x). \qquad (C.5)$$

The appropriate initial conditions are apparent from (C.3); they are $g(0) = 1$ and $g'(0) = 0$. With the usual ascending series expansion of Y_0 [51, Entries 9.1.13 and 9.1.12] and with some algebra, one can now verify that the answer in (6.5) leads to an ascending series for $g(x)$ that satisfies both the differential equation (C.5) and the initial conditions.

Appendix D: Additional Examples from the Electromagnetics Area

This appendix presents two additional integrals that can be evaluated using the Mellin-transform method. Apart from a complication arising in the second example when attempting to close the contour at left, both are straightforward.

D.1 AN INTEGRAL ARISING IN A RAIN ATTENUATION PROBLEM

In the study [24] of a rain attenuation problem, it is shown that

$$\int_0^1 \frac{e^{-xy} - 1}{y} (\ln y)^2 \, dy = -2x \, {}_4F_4(1, 1, 1, 1; 2, 2, 2, 2; -x), \qquad x > 0, \qquad (D.1)$$

an equality which we verify here by the Mellin-transform method. If $xf(x)$ denotes the left-hand side of (D.1), then

$$f(x) = (g \oslash h)(x), \qquad (D.2)$$

where

$$g(x) = \frac{e^{-x} - 1}{x} \qquad (D.3)$$

and

$$h(x) = \begin{cases} (\ln x)^2, & \text{if } 0 < x < 1, \\ 0, & \text{if } x > 1. \end{cases} \qquad (D.4)$$

The Mellin transform of $\tilde{g}(z)$ of $g(x)$ can be found by using (2.28) together with the identity (2.7) for $\alpha = -1$. It is

$$\tilde{g}(z) = \Gamma(z - 1), \qquad 0 < \text{Re}\{z\} < 1, \qquad (D.5)$$

whereas the Mellin transform $\tilde{h}(z)$ of $h(x)$ is provided in [6, Entry 8.4.6.3]

$$\tilde{h}(z) = 2 \left[\frac{\Gamma(z)}{\Gamma(z + 1)} \right]^3 = \frac{2}{z^3}. \qquad (D.6)$$

The second form in (D.6), which is simpler, follows from the first form by the recurrence formula (2.16). But, as explained in Section 2.8, the first form is preferable for the present purposes because it is a standard product. From (4.2), we get the Mellin–Barnes integral representation

$$f(x) = \frac{1}{2\pi i}\, 2 \int_{\delta-i\infty}^{\delta+i\infty} \frac{\Gamma(z-1)[\Gamma(1-z)]^3}{[\Gamma(2-z)]^3} x^{-z}\, dz, \qquad 0 < \delta < 1. \tag{D.7}$$

The quantity Δ defined in Chapter 8 is positive so that the contour can be closed at left and (4.3) becomes

$$f(x) = 2 \sum_{n=0}^{\infty} \mathrm{Res} \left\{ \frac{\Gamma(z-1)[\Gamma(1-z)]^3}{[\Gamma(2-z)]^3} x^{-z}; z = -n \right\}. \tag{D.8}$$

The poles to the left of the contour are due to $\Gamma(z-1)$, and the residues can be calculated with the aid of Application 3 of Section 2.7. We obtain

$$f(x) = 2 \sum_{n=0}^{\infty} \left[\frac{\Gamma(1+n)}{\Gamma(2+n)} \right]^3 \frac{(-1)^{n+1}}{\Gamma(n+2)} x^n, \tag{D.9}$$

from which the desired result (D.1) follows easily.

Note that it is very simple to obtain (D.1) directly by expanding $[\exp(-xy) - 1]/y$ into its Taylor series about the point $y = 0$ and then integrating term by term.

D.2   AN INTEGRAL RELEVANT TO THE THIN-WIRE LOOP ANTENNA

A number of works [66]–[69] have dealt with the problem of solvability of the usual (Hallén and Pocklington) integral equations for the current $I(z)$ on a straight, thin-wire antenna. For the case where the usual approximate (also called reduced) kernel is used, it is shown in [66]–[69] that many of the usual integral equations have no solution. That is, no function $I(z)$ can satisfy those equations (to make this statement precise, one must specify proper admissibility conditions on $I(z)$). We stress that the approximate kernel—as opposed to the so-called exact kernel—is a nonsingular function.

Recent work [52] extends the above considerations to certain popular integral equations, with nonsingular kernels, for the current on a thin circular loop antenna. The method consists of finding a formal solution of the integral equations using Fourier series and examining the convergence of the formal series. During the course of these investigations, an important step is to show that

$$f(x) = \int_0^\infty K_0(y) J_{2n}(xy) \, dy = \frac{1}{2} x^{2n} \frac{\left[\Gamma\left(n + \frac{1}{2}\right)\right]^2}{\Gamma(2n + 1)}$$

$$\times \, _2F_1\left(n + \frac{1}{2}, n + \frac{1}{2}; 2n + 1; -x^2\right), \qquad x > 2, \quad n = 0, 1, \ldots. \quad (D.10)$$

In (D.10), $x = 2b/a$, where b is the loop radius and a is the wire radius, so that $x > 2$. Although (D.10) can be deduced from integral tables (see [5, Entry 2.16.21.1]), we can verify (D.10) using the Mellin-transform method as follows.

The integral $f(x)$ can be written as in (4.1) where $g(x) = J_{2n}(x)$ and $h(x) = K_0(x)$. The Mellin transform $\tilde{g}(z)$ can be found in Table 2.1, while from [6, Entry 8.4.23.1] it follows that

$$\tilde{h}(z) = \frac{1}{4}\left(\frac{1}{2}\right)^{-z}\left[\Gamma\left(\frac{z}{2}\right)\right]^2, \qquad \text{Re}\{z\} > 0. \qquad (D.11)$$

We thus obtain the Mellin–Barnes integral representation

$$f(x) = \frac{1}{2\pi i} \int_{\delta - i\infty}^{\delta + i\infty} \frac{1}{2}\left[\Gamma\left(\frac{1}{2} - z\right)\right]^2 \frac{\Gamma(n + z)}{\Gamma(1 + n - z)} x^{-2z} \, dz, \qquad 0 < \delta < \frac{1}{2}, \qquad (D.12)$$

in which a change of variable was made so that all coefficients of z in the gamma functions are 1 or -1. There are simple poles to the left of the contour (contributed by $\Gamma(n + z)$) and double poles to the right of the contour, contributed by $[\Gamma(1/2 - z)]^2$. Furthermore, the parameters Δ and Π in Chapter 8 are $\Delta = 0$ and $\Pi = -2\ln x$. Thus in this example, in which $x > 2$, we cannot close the contour at left. But there are (at least) two ways to proceed:

(i) For all $x > 0$, $f(x)$ is an analytic function of x. We thus temporarily assume that $0 < x < 1$ so that, by the discussion in Chapter 8, the contour can be closed at left. Calculating the residues leads to a series representation, which can then be identified with a $_2F_1$:

$$f(x) = \frac{1}{2} x^{2n} \frac{\left[\Gamma\left(n + \frac{1}{2}\right)\right]^2}{\Gamma(2n + 1)} \, _2F_1\left(n + \frac{1}{2}, n + \frac{1}{2}; 2n + 1; -x^2\right), \qquad 0 < x < 1.$$
$$(D.13)$$

By analytic continuation (see Section 3.1 for the analytic continuation of the $_2F_1$), the result (D.13) can also be extended to all $x > 0$ (and, in particular, to $x > 2$), giving (D.10).

(ii) Alternatively, express the result in (D.12) as a G-function by comparing with (3.2). For all $x > 0$, this gives

$$f(x) = \frac{1}{2} G^{12}_{22} \left(x^2 \left| \begin{array}{cc} \frac{1}{2} & \frac{1}{2} \\ n & -n \end{array} \right. \right),$$

(D.14)

which can be simplified to (D.10) with the aid of the usual simplification tables. (Note that closing the contour at left when $0 < x < 1$ can also be justified on the grounds of Case G1 and Case G5 of Section 3.1.)

Bibliography

[1] G. Fikioris, "Integral evaluation using the Mellin transform and generalized hypergeometric functions: Tutorial and applications to antenna problems," *IEEE Trans. Antennas Propagat.*, vol. 54, no. 12, pp. 3895–3907, December 2006. doi:10.1109/TAP.2006.886579

[2] S. Wolfram, *The Mathematica Book*, 5th ed. Cambridge, UK: Wolfram Media, Inc., 2003, §A.10.3, §A.9.5.

[3] A. P. Prudnikov, Yu. A. Brychkov, and O. I. Marichev, "Evaluation of integrals and the Mellin transform," *J. Sov. Math.*, vol. 54, no. 6, pp. 1239–1341, May 1991. (Translated from *It. Nauk. Tekhniki, Ser. Mat. Anal.*, vol. 27, pp. 3–146, 1989.) doi:10.1007/BF01373648

[4] A. P. Prudnikov, Yu. A. Brychkov, and O. I. Marichev, *Integrals and Series: Elementary Functions*, vol. 1. Amsterdam, The Netherlands: Gordon and Breach, 1986.

[5] A. P. Prudnikov, Yu. A. Brychkov, and O. I. Marichev, *Integrals and Series: Special Functions*, vol. 2. London, UK: Taylor and Francis, 2002. (Reprint of 1986 ed.)

[6] A. P. Prudnikov, Yu. A. Brychkov, and O. I. Marichev, *Integrals and Series: More Special Functions*, vol. 3. London, UK: Taylor and Francis, 2002. (Reprint of 1990 ed.)

[7] O. I. Marichev, *Handbook of Integral Transforms of Higher Transcendental Functions: Theory and Algorithmic Tables.* New York: John Wiley & Sons, 1983.

[8] V. S. Adamchik and O. I. Marichev, "The algorithm for calculating integrals of hypergeometric type functions and its realization in REDUCE system," in *Proceedings of the Conference of ISSAC'90*, Tokyo, Japan, 1990, pp. 212–224.

[9] V. Adamchik, "Definite integration in *Mathematica* V3.0," *Mathematica in Education and Research*, vol. 5, no. 3, pp. 16–22, 1996. (Available online at http://www-2.cs.cmu.edu/~adamchik/articles/mier.htm).

[10] D. Zwillinger, *Handbook of Integration.* London, UK: Jones and Bartlett Publishers, 1992.

[11] M. J. Ablowitz and A. S. Fokas, *Complex Variables: Introduction and Applications, 2nd ed.* Cambridge, UK: Cambridge University Press, 2003.

[12] J. Bertrand, P. Bertrand, and J.-P. Ovarlez, "The Mellin transform," *The Transforms and Applications Handbook*, 2nd ed., A. D. Poularikas, Ed. Boca Raton, FL: CRC Press, Chapt. 11, 1999.

[13] R. J. Sasiela, *Electromagnetic Wave Propagation in Turbulence: Evaluation and Application of Mellin Transforms*. New York: Springer-Verlag, 1994.

[14] R. J. Sasiela and J. D. Shelton, "Mellin transform methods applied to integral evaluation: Taylor series and asymptotic approximations," *J. Math. Phys.*, vol. 34, no. 6, pp. 2572–2617, June 1993. doi:10.1063/1.530086

[15] G. F. Carrier, M. Krook, and C. E. Pearson, *Functions of a Complex Variable; Theory and Technique*. New York: McGraw-Hill, 1966.

[16] A. Erdélyi, W. Magnus, F. Oberhettinger, and F. G. Tricomi, *Higher Transcendental Functions*, vol. I. Malabar, FL: Krieger Publishing Co., 1981. (Reprint of 1953 ed.)

[17] A. Erdélyi, W. Magnus, F. Oberhettinger, and F. G. Tricomi, *Tables of Integral Transforms*, vol. 1. New York: McGraw-Hill, Chapt. VI, 1954.

[18] F. Oberhettinger, *Tables of Mellin Transforms*. New York: Springer-Verlag, 1974.

[19] I. S. Gradshteyn and I. M. Ryzhik, *Table of Integrals, Series, and Products*, 6th ed., Boston: Academic Press, 2000.

[20] http://functions.wolfram.com/HypergeometricFunctions/

[21] A. M. Mathai, *A Handbook of Generalized Special Functions for Statistical and Physical Sciences*. Oxford, UK: Clarendon Press, 1993.

[22] C. A. Balanis, *Antenna Theory: Analysis and Design*, 3rd ed. New York: Wiley, 2005.

[23] R. F. Harrington, *Time-Harmonic Electromagnetic Fields*. New York: McGraw-Hill, 1961, Section 4.11.

[24] A. D. Panagopoulos, G. Fikioris, and J. D. Kanellopoulos, "Rain attenuation power spectrum of a slant path," *Electron. Lett.*, vol. 38, no. 20, pp. 1220–1222, September 2002. doi:10.1049/el:20020815

[25] A. G. Derneryd, "Analysis of the microstrip disk antenna element," *IEEE Trans. Antennas Propagat.*, vol. AP-27, no. 5, pp. 660–664, September 1979. doi:10.1109/TAP.1979.1142159

[26] I. J. Bahl and P. Bhartia, *Microstrip Antennas*. Dedham, MA: Artech House, 1980, p. 93.

[27] J. D. Mahony, "Approximate expressions for the directivity of a circular microstrip-patch antenna," *IEEE Antennas and Propagation Magazine*, vol. 43, no. 4, pp. 88–90, August 2001. doi:10.1109/74.951561

[28] J. D. Mahony, "Correction," *IEEE Antennas and Propagation Magazine*, vol. 43, no. 5, p. 93, October 2001. doi:10.1109/MAP.2001.979376

[29] K. Verma and Nasimuddin, "Simple expressions for the directivity of a circular microstrip antenna," *IEEE Antennas Propagat. Mag.*, vol. 44, no. 5, pp. 91–95, October 2002. doi:10.1109/MAP.2002.1077780

[30] S. V. Savov, "An efficient solution of a class of integrals arising in antenna the-
      ory,"*IEEE Antennas Propagat. Mag.*, vol. 44, no. 5, pp. 98–101, October 2002.
      doi:10.1109/MAP.2002.1077781

[31] J. D. Mahony, "Circular microstip-patch directivity revisited: An easily computable exact
      expression," *IEEE Antennas Propagat. Mag.*, vol. 45, no. 1, pp. 120–122, February 2003.

[32] S. Adachi, "Correspondence," *IEEE Antennas Propagat. Mag.*, vol. 45, no. 1, p. 122,
      February 2003.

[33] S. Savov, "A comment on the radiation resistance," *IEEE Antennas Propagat. Mag.*,
      vol. 45, no. 3, p. 129, June 2003. doi:10.1109/MAP.2003.1232170

[34] J. D. Mahony, "A comment on Q-type integrals and their use in expressions for ra-
      diated power,"*IEEE Antennas Propagat. Mag.*, vol. 45, no. 3, pp. 127–128, June 2003.
      doi:10.1109/MAP.2003.1232169

[35] B. J. Stoyanov and R. A. Farrell, "On the asymptotic evaluation of $\int_0^{\pi/2} J_0^2(\lambda \sin x)\, dx$,"
      Math. Comput., vol. 49, pp. 275–279, 1987. doi:10.2307/2008265

[36] R. Wong, "Asymptotic expansion of $\int_0^{\pi/2} J_v^2(\lambda \cos \theta)\, d\theta$," *Math. Comput.*, vol. 50,
      pp. 229–234, 1988. doi:10.2307/2007926

[37] B. J. Stoyanov, R. A. Farrell, and J. F. Bird, "Asymptotic expansions of integrals of two
      Bessel functions via the generalized hypergeometric and Meijer functions," *J. Comput.
      Appl. Math.*, vol. 50, pp. 533–543, 1994.

[38] D. Margetis and G. Fikioris, "Two-dimensional, highly directive currents on large
      circular loops," *J. Math. Phys.*, vol. 41, no. 9, pp. 6130–6172, September 2000.
      doi:10.1063/1.1288245

[39] B. J. Stoyanov, "Comment on 'Two-dimensional, highly directive currents on large
      circular loops,' [J. Math. Phys. 41, 6130 (2000)]," *J. Math. Phys.*, vol. 45, no. 4,
      pp. 1711–1712, April 2004. doi:10.1063/1.1650536

[40] L. J. Landau and N. J. Luswili, "Asymptotic expansion of a Bessel function integral
      using hypergeometric functions," *J. Comput. Appl. Math.*, vol. 132, pp. 387–397, 2001.
      doi:10.1016/S0377-0427(00)00441-6

[41] B. J. Stoyanov, "Comments on 'Asymptotic expansion of a Bessel function integral using
      hypergeometric functions' by L. J. Landau and N. J. Luswili," *J. Comput. Appl. Math.*,
      vol. 176, pp. 259–262, 2005. doi:10.1016/j.cam.2004.07.015

[42] G. Fikioris, P. G. Cottis, and A. D. Panagopoulos, "On an integral related to bi-
      axially anisotropic media," *J. Comput. Appl. Math.*, vol. 146, pp. 343–360, 2002.
      doi:10.1016/S0377-0427(02)00368-0

[43] `http://functions.wolfram.com/HypergeometricFunctions/`, see the page
      "Asymptotic series expansions for $q = p + 1$."

[44] S. Toumpis, "Radiation of elementary dipole in infinite biaxial, anisotropic medium," (in Greek), Senior thesis, Greece. Athens: National Technical University of Athens, July 1997.

[45] P. G. Cottis, private communication, 1997.

[46] P. G. Cottis and G. D. Kondylis, "Properties of the dyadic Green's function for an unbounded biaxial medium," *IEEE Trans. Antennas Propagat.*, vol. 43, no. 2, pp. 154–161, February 1995. doi:10.1109/8.366377

[47] P. G. Cottis, C. N. Vazouras, and C. Spyrou, "Green's function for an unbounded biaxial medium in cylindrical coordinates," *IEEE Trans. Antennas Propagat.*, vol. 47, no. 1, pp. 195–199, January 1999. doi:10.1109/8.753010

[48] A. Sommerfeld, *Partial Differential Equations in Physics.* New York: Academic Press, 1959, pp. 188–200.

[49] S. W. Lee and Y. T. Lo, "Current distribution and input admittance of an infinite cylindrical antenna in anisotropic plasma," *IEEE Trans. Antennas Propagat.*, vol. 15, no. 2, pp. 244–252, March 1967. doi:10.1109/TAP.1967.1138871

[50] G. Fikioris, "Table Erratum 634," *Math. Comput.*, vol. 67, pp. 1753–1754, October 1998.

[51] I. Abramowitz and I. A. Stegun, Eds., *Handbook of Mathematical Functions with Formulas, Graphs, and Mathematical Tables (National Bureau of Standards Applied Mathematics Series, Vol. 55).* Washington, D.C.: U. S. Government Printing Office, 1972.

[52] G. Fikioris, P. J. Papakanellos, and H. T. Anastassiu, "On the use of nonsingular kernels in certain integral equations for thin-wire circular-loop antennas," *IEEE Trans. Antennas Propagat.*, submitted.

[53] B. Davies, *Integral Transforms and Their Applications*, 2nd ed. New York: Springer-Verlag, 1985.

[54] I. N. Sneddon, *The Use of Integral Transforms,* New York: McGraw-Hill, 1972.

[55] E. T. Whittaker and G. N. Watson, *A Course in Modern Analysis*, 4th ed. New York: Cambridge University Press, 1927 (reprinted, 1992).

[56] Y. L. Luke, *The Special Functions and Their Approximations*, vols. I and II. New York: Academic Press, 1969.

[57] K. Roach, "Meijer G function representations," in *Proceedings of the 1997 International Symposium on Symbolic and Algebraic Computation.* New York: ACM Press, 1997, pp. 205–211.

[58] R. B. Paris and D. Kaminski, *Asymptotics and Mellin–Barnes Integrals.* Cambridge, UK: Cambridge University Press, 2001.

[59] A. M. Mathai and R. K. Saxena, *The H-Function with Applications in Statistics and Other Disciplines.* New Delhi, India: Wiley Eastern, Ltd., 1978.

[60] R. Wong, *Asymptotic Approximations of Integrals.* New York: Academic Press, 1989.

[61] F. W. J. Olver, *Asymptotics and Special Functions*. Natick, MA: A. K. Peters, Ltd, 1997.

[62] N. Bleistein and R. A. Handelsman, *Asymptotic Expansions of Integrals*. New York: Dover, 1986.

[63] F. Mainardi and G. Pagnini, "Salvatore Pincherle: the pioneer of the Mellin–Barnes integrals," *J. Computat. Appl. Math.*, vol. 153, pp. 331–342, 2003.

[64] D. V. Widder, *Advanced Calculus*, 2nd ed. Englewood Cliffs, NJ: Prentice-Hall, 1961 (reprinted, New York: Dover, 1989), Chapt. 10.

[65] M. A. Evgrafov, "Series and integral representations," in *Analysis, I*. R. V. Gamkrelidze, Ed. New York: Springer, Chapt. I, 1989. (Translated from the Russian by D. Newton).

[66] T. T. Wu, "Introduction to linear antennas," in *Antenna Theory*, Pt. I. R. E. Collin and F. J. Zucker, Eds. New York: McGraw-Hill, Chapt. 8, 1969.

[67] G. Fikioris and T. T. Wu, "On the application of numerical methods to Hallén's equation," *IEEE Trans. Antennas Propagat.*, vol. 49, no. 3, pp. 383–392, March 2001. doi:10.1109/8.918612

[68] G. Fikioris, "The approximate integral equation for a cylindrical scatterer has no solution," *J. Electromagn. Waves Appl.*, vol. 15, no. 9, pp. 1153–1159, September 2001.

[69] G. Fikioris, J. Lionas, and C. G. Lioutas, "The use of the frill generator in thin-wire integral equations," *IEEE Trans. Antennas Propagat.*, vol. 51, no. 8, pp. 1847–1854, August 2003. doi:10.1109/TAP.2003.815412

Author Biography

George Fikioris was born in Boston, MA, on December 3, 1962. He received the Diploma of Electrical Engineering from the National Technical University of Athens, Greece (NTUA), in 1986, and the S.M. and Ph.D. degrees in Engineering Sciences from Harvard University in 1987 and 1993, respectively. From 1993 to 1998, he was an electronics engineer with the Air Force Research Laboratory, Hanscom AFB, MA. From 1999 to 2002, he was a researcher with the Institute of Communication and Computer Systems at the NTUA. From 2002 to February 2007, he was a lecturer at the school of Electrical and Computer Engineering, NTUA. In February 2007, he became an assistant professor at that school. He is the author or coauthor of over 25 papers in technical journals and numerous papers in conferences. Together with R. W. P. King and R. B. Mack, he has coauthored *Cylindrical Antennas and Arrays*, Cambridge University Press, 2002. His research interests include electromagnetics, antennas, and applied mathematics.

Dr. Fikioris is a senior member of the IEEE (Antennas & Propagation, Microwave Theory & Techniques, and Education Societies), and a member of the American Mathematical Society and of the Technical Chamber of Greece.